T0172791

THE LIQUID CRYSTALS BOOK SERIES

NANOSTRUCTURES AND NANOCONSTRUCTIONS BASED ON DNA

THE LIQUID CRYSTALS BOOK SERIES

Edited by
Virgil Percec
Department of Chemistry
University of Pennsylvania
Philadelphia, PA

The Liquid Crystals book series publishes authoritative accounts of all aspects of the field, ranging from the basic fundamentals to the forefront of research; from the physics of liquid crystals to their chemical and biological properties; and from their self-assembling structures to their applications in devices. The series will provide readers new to liquid crystals with a firm grounding in the subject, while experienced scientists and liquid crystallographers will find that the series is an indispensable resource.

PUBLISHED TITLES

THE LIQUID CRYSTALS BOOK SERIES

NANOSTRUCTURES AND NANOCONSTRUCTIONS BASED ON DNA

Yuri M. Yevdokimov
V. I. Salyanov
S.G. Skuridin

CRC Press
Taylor & Francis Group
Boca Raton   London   New York

CRC Press is an imprint of the
Taylor & Francis Group, an **informa** business

CRC Press
Taylor & Francis Group
6000 Broken Sound Parkway NW, Suite 300
Boca Raton, FL 33487-2742

First issued in paperback 2019

© 2012 by Taylor & Francis Group, LLC
CRC Press is an imprint of Taylor & Francis Group, an Informa business

No claim to original U.S. Government works

ISBN-13: 978-1-4665-0569-8 (hbk)
ISBN-13: 978-0-367-38129-5 (pbk)

Visit the Taylor & Francis Web site at
http://www.taylorandfrancis.com

and the CRC Press Web site at
http://www.crcpress.com

Contents

Foreword

The possibility of creation of nanostructures as a result of a so-called hybridization technology, based on self-assembly of the DNA molecules, was substantiated in pioneering works by N. C. Seeman, a U.S. scientist, in the early 1990s.

The hybridization technology of self-assembly of DNA nanostructures is based on a simple principle: Two flexible single-stranded DNA fragments (molecules) must form a rigid (in nanometer scale) double-stranded fragment (molecule) if the sequences of nitrogen bases of the initial fragments are complementary. Therefore, the character of interaction between the initial single-stranded molecules and the type of a "flat lattice" formed by double-stranded DNA fragments cannot only be predicted, but controlled as well by choosing the corresponding sequences of nitrogen bases. However, the flexibility of the obtained lattices restricted the creation of large-scale (extended) two-dimensional nanostructures. This stimulated further research and synthesis of new initial DNA fragments with higher stiffness.

At the same time, when forming DNA nanostructures by the hybridization technique, the goal is not only to synthesize various two- and three-dimensional structures, but also to create functional devices that could be used in practice. Despite the introduction of cruciform structures, the Holliday structures, double- and triple-crossover DNA molecules, DNA tiles, etc., into experimental practice, the creation of large-scale two- and three-dimensional DNA nanostructures is still a costly and complicated task.

One can add that a number of limitations to the use of the hybridization technique—such as a low yield of tailored products (a few percent) and the slow rate of a precise self-assembly, which can take several days—have a negative influence on the practical application of the obtained DNA nanostructures.

In 1994, for the first time, a supposition was formulated, according to which (in contrast to the hybridization approach) adjacent double-stranded DNA molecules in the content of ordered layered structures could be used as building blocks for creation of spatial constructions. Such layered structures, in essence, represent by themselves the DNA liquid-crystalline structures. Indeed, the formation of DNA liquid-crystalline structures takes place spontaneously, at room temperature, under strictly controllable conditions with a yield of approximately 90%–95%.

The properties of DNA liquid crystals and DNA liquid-crystalline dispersions—as well as the methods of research of these properties—are well known by the physicists and chemists dealing with polymeric liquid crystals, including the author's team, and they are well described and documented (see, for instance, *DNA Liquid-Crystalline Dispersions and Nanoconstructions*, CRC Press, Boca Raton, FL 2011).

This current book, *Nanostructures and Nanoconstructions Based on DNA*, explores three important interrelated issues actively discussed by scientists of various specialties both in Russia and abroad.

First, the authors have made an attempt to show that nanoobjects are not only described by their size, but also (and mostly) by a number of specific physical and chemical properties that depend on their size. They stress the special significance of biological molecules to the formation of new branches of nanotechnology. A great deal of attention is paid to the issue of the toxicity of nanoparticles and nanostructures of various origins.

Second, based on the properties of biological molecules, namely DNA, the principal approaches to the hybridization assembly of various planar DNA nanostructures are minutely analyzed and compared, and the peculiarities and disadvantages of the created DNA nanostructures are enumerated.

Finally, a great deal of attention is paid to the creation of spatial nanoconstructions based on DNA, i.e., to the formation of three-dimensional constructions, using as building blocks spatially ordered, neighboring double-stranded DNA molecules. Both the basic principles of obtaining various DNA liquid-crystal dispersion particles and the modern methods of their research are considered here. It is especially important to note that the mutual orientation of DNA fragments in the layers formed by these fragments can be deliberately adjusted (changed). The "rigid" spatial nanoconstructions of DNA obtained by the authors have attracted great interest due to their unique properties. It is possible that future developments—choosing nitrogen base sequences and other factors that affect the secondary structure of DNA in neighboring layers; manipulation of the rigid nanoconstructions, such as cutting them and moving them under controllable conditions; and further enzyme processing—will lead to the creation of a practically significant pattern from fragments of adjacent DNA molecules or even a multicored nanometer wire that may be used in various devices.

While deliberating upon these issues, the authors state a very important idea; namely, they show that the equations that describe the dimensional properties of metallic nanoparticles are quite suitable for nanoobjects formed by nitrogen bases as well as DNA nanostructures and nanoconstructions, i.e., they point out that in the area of nanoobjects, physical laws are basically versatile.

The possible practical applications of DNA nanoconstructions with various properties, for instance, as biosensing units for optical sensory systems capable of detecting chemically or biologically active compounds, are described in the final chapter of the book.

One can hope that the readers of this book specializing in different areas of physics and chemistry will not only find statements that can be criticized, but also discover multiple possibilities to apply their own knowledge to form DNA nanostructures and nanoconstructions, and especially to use such structures in practice. In this case, it can be expected that the interaction of physicists, chemists, and biologists researching different properties of DNA molecules and interested in the exploration of biological processes will lead to new results useful for nanotechnology, biotechnology, and practical medicine.

A. I. Grigoriev

Editor's Preface

To the present time, a number of monographs written by Russian scientists on nanotechnology have been published:

Gusev, A. I. 2005. "Nanomaterials, nanostructures and nanotechnology." Moscow: Fizmatlit.

Menshutina, N. V. 2006. "Introduction to nanotechnology." Kaluga, Russia: Science Literature Publishing House of N. F. Bochkareva.

Suzdalev, I. P. 2006. "Physical chemistry of nanoclusters, nanostructures and nanomaterials." Moscow: KomKniga.

Rambidi, N. G., and A. V. Berezkin. 2008. "Physical and chemical basis of nanotechnology." Moscow: Fizmatlit.

As well, a number of books written by foreign scientists and edited by leading Russian nanotechnology specialists have been translated into Russian:

Hartmann, U. 2008. *Faszination Nanotechnologie*, edited by L. N. Patrikeev. Moscow: Binom.

Kobayasi, N. 2008. *NanoTechnology*, edited by L. N. Patrikeev. Moscow: Binom.

In addition, a collection of articles from 2006:

Nanomaterials. Nanotechnology. Nanosystem Equipment, edited by P. P. Maltsev. Moscow: Technosfera

as well as a book from 2005:

Introduction to Nanotechnology, by C. P. Poole and F. J. Owens. Moscow: Technosfera

were published in the series *The World of Materials and Technologies*.

However, only one book (from 2007) is more or less devoted to the description of specific details of the creation of nanostructures from biological molecules:

Nanotechnology, edited by C. M. Niemeyer and C. A. Mirkin. Wiley-VCH, 5th reprint.

Nevertheless, the interest in nanostructures (nanoobjects) created from biological molecules, namely DNA, is increasing every year. This is proved by the increasing number of scientific articles focused on this issue published in various journals. Moreover, the discussed possibilities of practical application of DNA nanoobjects indicate that the creation of such objects not only reflects the intellectual power of scientific teams, but also is driven by commercial interests.

The book *Nanostructures and Nanoconstructions Based on DNA*, as follows from its title, is devoted to the two specific problems related to DNA molecules. At the same time, it explores other questions that may be interesting to scientists from various institutes working in the area of creating nanostructures based on biological molecules and the research of their properties. These issues include:

Properties of nanoparticles formed by biological molecules
Toxicity of nanoparticles
Questions related to the practical application of DNA-based nanoobjects

This book presents a compilation of actual experimental results supported by multiple illustrations, which makes it possible to form a precise concept of the techniques of formation of various DNA-based nanostructures and nanoconstructions as well as the properties of these objects. This information opens a gateway to the practical application of these structures.

The authors did not consider the modern physical methods of research of all nanoobjects, though they have illustrated the results of application of the enumerated methods to explore the properties of DNA nanoconstructions. At the same time, the authors have not mentioned the results received in several areas of science that nowadays may be indirectly related to DNA nanostructures and nanoconstructions. Specifically, I refer to the conductive properties of DNA; the properties of so-called metal-ion-modified DNA (metallic DNA or M-DNA); DNA-nanowires, where DNA molecules are used as templates for deposition of various conductive metals; the creation of motors or molecular walking devices based on DNA; as well as the computing procedures realized with the use of DNA molecules. Many of the results received in these areas are still subject to discussion among scientists in different countries, and they are not yet definite and convincing enough to be presented here. It is possible that these problems will become subjects of future activity of the authors.

A few notes regarding the processing of this book, *Nanostructures and Nanoconstructions Based on DNA*, may be helpful. While analyzing the materials presented in the original Russian version of our book and evaluating its scientific merits, the publisher's reviewer (CRC Press) formulated an interesting statement:

When one talks about the nanostructures and nanoconstructions based on DNA, one should, of course, devote some time also to the quintessential nanoobject associated with DNA: the viruses. There are other books on the molecular biology of viruses, but I do believe that the authors should include the connection between viral packing of DNA and its formation of ordered phases in the bulk into one of the chapters, or preferably introduce it in a chapter of their own. This might not be part of the concept they envision for the book, but I would like to propose this extension, which would not only help the general scope of the book, but help it attract additional readership.

The authors accepted this as a very interesting statement and subsequently introduced certain modifications into the content and figures of this book. These modifications consist, mainly, in reducing the length of the theoretical descriptions of various properties of the DNA liquid-crystalline dispersions.

Here I want to add the following: The authors have previously analyzed the peculiarities of DNA packing in the viral heads and in the protozoan chromosomes in their first book, *DNA Liquid-Crystalline Dispersions and Nanoconstructions* (2011, CRC Press), and one can direct readers interested in this issue to their previous book. However, after discussion with the editor's team, the authors have attempted to add a new chapter dedicated to rigid biological nanoconstructions to the English version of the book, *Nanostructures and Nanoconstructions Based on DNA*. For this purpose, the authors invited a Russian team—the famous specialists from the Gamaleya Institute of Epidemiology and Microbiology of the Ministry of Health (Moscow, Russia)—to take part in the writing of Chapter 5. Hence, in contrast to the Russian version, the English variant of this book contains both modified content and an additional chapter.

Readers may feel a need to comment on or criticize some parts of the book, perhaps demanding a more detailed description. Any comments made by the readers shall be accepted by the authors with gratitude and considered in their further work.

Yu. M. Yevdokimov

Authors' Preface

After publication in Russia of our book *DNA Liquid-Crystalline Dispersions and Nanoconstructions*, colleagues from various scientific organizations not only made a number of stimulating remarks about it, but also brought our attention to the necessity for a more detailed description of the materials on a number of issues, specifically, a more detailed description of the properties of nanoobjects formed by biopolymeric molecules and, more importantly, a comparison of different laboratory technologies used by the authors from famous laboratories to create DNA nanostructures and nanoconstructions. Moreover, numerous conferences and seminars on various aspects of nanotechnology that took place in Moscow have shown that these very issues attract the interest of scientists who work in biological institutes and seek to apply their knowledge in the area of nanotechnology. We have tried to take into account the most interesting remarks and wishes stated by our colleagues in this book, *Nanostructures and Nanoconstructions Based on DNA*. The book reflects our point of view that had been formed at the time of its writing.

As authors, we understand that the book, unlike the rapidly developing DNA nanotechnology, is complete. That is why the book represents by itself a snapshot of the branches of DNA structural nanotechnology. The fact that new results may appear in this area of science tomorrow is not a reason to waive the analysis of the content of this snapshot, which nevertheless reflects the principal tendencies in the development of DNA structural nanotechnology.

We would like to thank our colleagues from various laboratories and organizations, namely, V. S. Prassolov (Engelhardt Institute of Molecular Biology of the RAS), V. M. Rudoy (Frumkin Physical Chemistry and Electrochemistry Institute of the RAS), O. N. Kompanets (Institute of Spectroscopy of the RAS), V. V. Volkov (Shubnikiv Crystallography Institute of the RAS), S. V. Akulinichev (Institute of Nuclear Researches of the RAS), V. N. Nikiforov (Lomonosov Moscow State University), P. Laggner (AAS Biophysics and Nanosystem Research Institute, Graz, Austria), S. A. Saunin (AIST-NT Company, Zelenograd, Russia), and their coworkers not only for stimulating discussions, but also for the opportunity to reproduce some of the unpublished results. For technical reasons, we cannot enumerate all of the participants at all stages of the work, but their names can easily be found in the references to each chapter of this book.

We would like to express special gratitude to Professor N. C. Seeman (New York University) for interesting discussions and the gracious permission to reproduce some of the colored figures from his articles published in various scientific journals, quoted mostly in Chapter 2. We adapted these figures to the content of Chapter 2, and we use them with attribution: (Courtesy of Nadrian C. Seeman).

We want to express our cordial thanks to our colleagues: Professors D. Yu. Logunov, B. S. Naroditsky, and A. L. Gintsburg from Gamaleya Institute of Epidemiology and Microbiology of the Ministry of Health (Moscow, Russia) for their kind agreement to write an additional chapter, Chapter 5, in our book. This additional chapter was included in an attempt to attract the attention of the scientific audience to nanotechnology based on rigid viral particles.

We are thankful to the New Energetical Technologies Co. (Moscow, Russia) for financial support during the preparation and publication of the Russian version of our book. A minor part of the new results reported in Chapter 6 was obtained in the framework of State Contract No. 14.527.12.0012.

We are also grateful to our wives for creating the conditions for our productive work that allowed us to present the scientific material included in this book.

Introduction

THE GENERAL CONCEPT OF NANOTECHNOLOGY

dove la natura finisce di produrre
le sue spezie, l'uomo quivi cominicia,
con le cose naturali, con l'aiutorjo
di essa natura, a creare infinite spezie

Leonardo da Vinci
Disegni Anatomici
The Royal Library of Windsor

The terms *nanotechnology, nanoparticles*, and *nanomaterials* are familiar to a broad audience of readers. Indeed, manipulations on the level of isolated atoms make it possible to create new structured materials and devices with predictable and unique properties.

In the focus of the new area of science that has appeared at the cross-point of physics, chemistry, electronics, and computer science—known as nanotechnology— that started developing rapidly at the turn of the twentieth and the twenty-first century are the so-called nanoparticles, or nanosized objects, ranging from less than one nanometer to hundreds (thousands) of nanometers. (*Nano* comes from the Greek word *nanos*—a "dwarf," the quantity of 10^{-9} meters [nanometer, nm], i.e., a quantity comparable to the size of a single atom.) In a broad sense, nanotechnology involves the "creation and application of nanomaterials, devices and technical systems whose functions are determined by the presence of nanostructural fragments" [1].

The term *nanotechnology* was first used by Prof. Norio Taniguchi in 1974 in his report at a conference in Tokyo [2, 3]. Though the term appeared in the 1970s, it does not mean that technologists and scientists had not been working intuitively in this area before. Master artists of ancient Rome empirically used gold nanoparticles to color goblets and other glass products [4]. (This technology was used to create the famous Roman ruby goblets, such as the goblet of Lycurgus [the fourth century] stored in the British museum.) It is also appropriate to remember that in 1857, M. Faraday obtained and was working with a red colloid solution that contained gold nanoparticles with a size of 20 nm.

The science of nanotechnology and nanoparticles is relatively new on the scale of modern science. It is based on inventions made at the end of the twentieth century. The first of these inventions belongs to Nobel laureates, physicists Gerd Binnig and Heinrich Rohrer from IBM research laboratory, who designed the tunnel scanning microscope that made it possible to see single atoms in 1981 [5]. The second discovery was made in 1986, when the atoms were not only observed, but also manipulated for the first time with the help of a tunnel microscope upgraded by G. Binnig. The Nobel lecture in 1986 [6] described both the principle of operation of the tunnel microscope and manipulating atoms and the prospects of this invention. The authors told about economic prospects of the companies that were going to produce items based on nanotechnology. In the opinion of the authors, the size and the volume of materials used in this kind of production could not be large, which meant that the production itself was going to be inexpensive.

The current conception of nanotechnology arose from the modern achievements and inventions in the area of visualization, manipulation, and analysis of nanometer-sized structures as well as the controllable creation of new functional materials with unique properties and nanosized devices that provided technological breakthroughs in various industrial fields.

Before evaluating the advantages and disadvantages of nanotechnology, it is reasonable to define the principal terms used in this rapidly developing branch of science. One of the first definitions of nanotechnology can be found in an American nanotechnology program known as the U.S. National Nanotechnology Initiative (NNI, 1999). Nanotechnology, as defined by NNI is "the understanding and control of matter at dimensions of roughly 1–100 nanometers, where unique phenomena enable novel applications."

The report of the Royal Society & the Royal Academy of Engineering of 2004 [7] gives a quite similar definition of nanotechnology: "Nanotechnology is the design, characterization, production and application of structures, devices and systems by controlling shape and size at nanometer scale."

The Concept of the Development of Research in the Area of Nanotechnology in Russia until 2010 (approved by the government of the Russian Federation on 18 October 2004) used the following definition: "Nanotechnology is the combination of methods and techniques that provide the possibility of controllable creation and modification of objects that include components with the size of less than 100 nm which have unique properties and can be integrated with complete functioning large-scale systems."

The work by Silva [8] provides a more precise definition: "Nanotechnologies are technologies that use engineered materials or devices with the smallest functional organization on the nanometer scale *in at least one dimension*, typically ranging from 1 to ≈100 nanometers."

The definition of nanotechnology given on the website of Russian Nanotechnology Corporation (2008) is: "Nanotechnology is the area of applied science and technology devoted to the research of the properties of objects and development of devices with nanometer scale size (10^{-9} nm)*." [Here, the asterisk mark (*) corresponds to the following interesting remark: "In practice, depending on the area of science and technology, the size both below 10 nm and below 500 nm can be considered nanometric."]

It is obvious that all of the definitions are not perfect and can be subject to criticism, and the rapid development of various branches of nanotechnology will correct and improve them.

The principal scheme of integral parts of nanotechnology formed by the end of the twentieth century is illustrated in Figure 1. Commenting on this scheme, it can

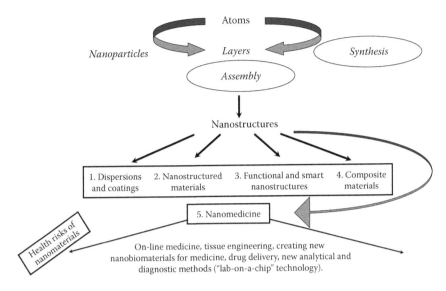

FIGURE 1 The principal scheme of the basic parts of nanotechnology formed to the end of the twentieth century and the main research fields.

be stated that engineering (technical) nanotechnology is directed toward solution of the following problems [9]:

- Creation of solid materials and surfaces with controllable molecular structures
- Designing new types of chemical compounds with controllable properties (nanoconstructions)
- Creating nanosized self-organizing or self-replicating structures
- Fabrication of devices for various purposes (components of nanoelectronics, nanooptics, nanoenergetics, etc.)
- Integration of nanosized devices with electric systems

According to the resolution of the 7th International Nanotechnology Conference (Wiesbaden, Germany), the scheme should be changed because nanoparticles, nanotubes and nanofibers, nanodispersions (colloids), nanostructured films and surfaces, nanoporous structures, nanocrystals, and nanoclusters have been attributed to nanomaterials.

Nanoparticles are an important state of a condensed phase. Nanoparticles can be considered an intermediate state between atoms and solid state. In the first approximation, nanoparticles can be defined as objects with at least one dimension varying from 1 nm to 100 nm containing atoms of one or several elements [10].

Metal clusters are multinuclear complex compounds based on a frame consisting of metal atoms and surrounded with ligands. A cluster is a nucleus that contains more than two atoms [11].

To create new nanomaterials, several approaches called "top-down" and "bottom-up" technologies are used (Figure 2).

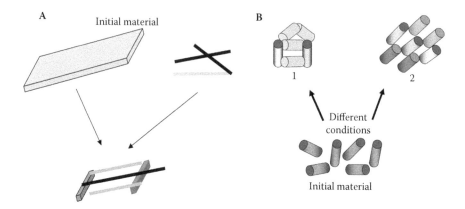

FIGURE 2 Technological approaches to the creation of nanomaterials: (A) top-down technology; (B) bottom-up technology.

In the case of top-down technologies (Figure 2(A)), nanomaterials (devices) are created by reducing the size of the initial substance or placement of required atoms on the surface of some material in a certain configuration. In particular, top-down technologies include different methods of lithography and atomic force microscopy.

In the case of bottom-up technologies (Figure 2(B)), nanomaterials (devices) are created from desired elements, such as atoms, that are integrated into spatial structures. Bottom-up techniques start with one or more objects of a certain kind integrated into various ordered (1, 2) structures as a result of a physicochemical (biochemical) process. An example of this approach is the process of formation of molecular ensembles (self-organization of molecules) that are triggered by the change in the physical or chemical properties of the medium. It is necessary to highlight that in any variant of the considered technologies the forming nanomaterials obtain new properties that the initial objects do not possess. This means that nanotechnologies cause, first of all, the appearance of new functional properties of the created objects. At the same time, though chemical or/and physical methods of the creation of nanomaterials (devices) are important in the whole process of nanotechnology, they are secondary to the functional properties of the created nanomaterials (devices).

REFERENCES

1. Tretyakov, Yu. D. 2007. Problems of development of nanotechnologies in Russia and abroad. *Herald of the Russian Academy of Sciences* (Russian ed.) 77:3–10.
2. Taniguchi, N. 1974. On the basic concept of nano-technology. *Proc. Intl. Conf. Prod. Eng. Tokyo, Part II, Jap. Soc. Pres. Eng.* 2:18–23.
3. Taniguchi, N. 1992. Future trends of nanotechnology. *Int. J. Japan. Soc. Prec. Eng.* 26:1–7.
4. Ratner, M., and D. Ratner. 2004. *Nanotechnology: a simple explanation of the next brilliant idea* (Russian ed.). Moscow: Ed. House Williams.
5. Binnig, G., H. Rohrer, Ch. Gerber, and E. Weibel. 1982. Surface studies by scanning tunneling microscopy. *Phys. Rev. Lett.* 49:57–60.

6. Binnig, G., and H. Rohrer. 1986. Scanning tunneling microscopy: From birth to adolescence. Nobel lecture, December 8.

7. 2004. Nanoscience and nanotechnologies: Opportunities and uncertainties. In *Royal Society and Royal Academy of Engineering*, 10. London: The Royal Society.

8. Silva, G. A. 2006. Neuroscience nanotechnology: Progress, opportunities and challenges. *Nature Rev. Neurosci.* 7:65–74.

9. Yevdokimov, Yu. M., M. A. Zakharov, and S. G. Skuridin. 2006. Nanotechnology based on nucleic acids. *Herald of the Russian Academy of Sciences* (Russian ed.) 76:112–20.

10. Sergeyev, G. B. 2007. *Nanochemistry: Textbook* (Russian ed.). Moscow: University—The Book House.

11. Suzdalev, I. P. 2006. Nanotechnology: The development and prospects. *Herald of the Russian Foundation for Basic Research* (Russian ed.) 6:27–46.

1 Nanoparticles and Biological Molecules

1.1 METAL NANOPARTICLES AND THEIR PROPERTIES

Starting with the pioneering works by H. Gleiter and his colleagues [1, 2], the interest in nanomaterials has increased rapidly because of their unique properties [3]. The created nanomaterials, namely, nanoparticles are on the edge of the nanotechnology wave. The current and potential applications of nanoparticles are broadening and covering a wide segment of the market. The analysis has shown that the market of nanomaterials reached $900 million in the United States in 2005 and is going to reach $11 billion by 2010 [4].

It should be noted that the transition from micro- to nanoparticles is accompanied by a qualitative change in many physicochemical properties of substances, such as melting and hardening temperature, solubility, pressure necessary for rearrangement of the lattice, kinetics of chemical processes on the surfaces of particles, etc. This is caused by the fact that, for the particles with at least one dimension comparable to (or smaller than) the correlation radius of a physical or chemical property (length of the electron-free track, size of germ of a new phase, size of the magnetic domain, etc.), the scale effects start to appear [5].

For metallic nanoparticles, two types of size effects are distinguished: the interior and the exterior effects. The interior effect is caused by the specific changes in the volumetric and spatial properties of both separate particles and ensembles obtained by self-organization of the particles. The exterior effect is a size-determined response to the exterior field or forces independent from the interior effect.

Some of the scale effects are detected at particle size below 10 micrometers, but these effects are especially distinctive in the case of nanoparticles. An example of an explicit size effect is the change in color of gold nanoparticles [6], which is illustrated in Figure 1.1. The figure contains electron microscopy images of particles (all scale bars are 100 nm). For spherical particles, the size varies from 4 to 40 nm (a–e); whereas for rods, the aspect ratio (the relation between the diameter, d, and the length, l) varies from 1.3 to 1.5 (for short rods, f–j) and reaches 20 for long rods (k). The change in color is more distinctive for the rods than for the spherical nanoparticles of gold. It follows from the shown images that as the size of the particles decreases, the color changes from brown to purple; moreover, the effect depends on the shape (globular or cylindrical) of the particles.

An analysis of the size effects in nanomaterials can be found in reviews published in 1999–2001 (see, for instance, [7, 8]). In material science, physics, chemistry, and biology, the effect of the size of grain (crystallite) on the properties of materials is a

1

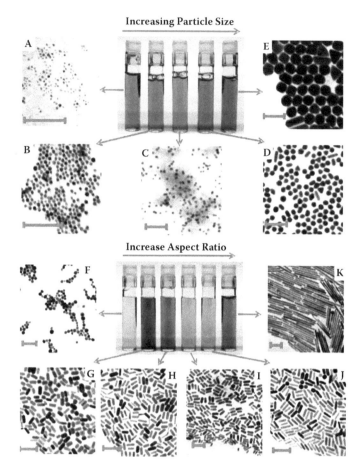

FIGURE 1.1 (See color insert) Change in color of water solutions of spherical gold nanoparticles (upper panels) and gold nanorods (lower panels), depending on the dimensions of the particles.

well-known problem. However, in the case of nanoparticles, there are at least several principal peculiarities of size effects.

1.1.1 PROPERTIES OF ATOMS ON THE SURFACE OF A PARTICLE DETERMINE THE APPEARANCE OF A NUMBER OF NEW PROPERTIES

Specific properties of nanoparticles are caused by the fact that the share of atoms on the surface of the particle is comparable to the total number of atoms in the particle. For example, for particles with a size of approximately 1 nm and a shape close to spherical, the ratio between spatial and interior atoms is close to 1. At the increase of d (to tens of nm and in some cases to several microns), the effect of this ratio on the properties of the substance is still quite strong even though the share of spatial atoms decreases. The atoms on the surface have fewer contacts with the adjacent atoms than the interior atoms, which makes them less stable.

The highly developed specific surface area of nanoparticles can be defined as

$$\Omega = K/d\rho$$

where K is a nondimensional coefficient ($K = 6$ for spheres and cubes) and ρ is the density of the disperse phase, which determines the excess Helmholtz energy (in comparison to macroscopic systems with analogous structure)

$$\Delta F_s = \sigma_{12}\Omega$$

where σ_{12} is the surface tension of the "nanoparticle–medium" interphase boundary.

The value of ΔF_s determines the intensity and the direction of processes that involve nanoparticles. Therefore, for nanoparticles with relatively low surface tension

$$\sigma_{12} \leq bk_B T/d^2 \tag{1.1}$$

where b is a numerical coefficient; k_B is the Boltzmann constant; and T is temperature. The values for σ_{12} are quite far from the thermodynamic equilibrium. Consequently, processes that cause reduction of surface energy can take place spontaneously in such systems. In particular, particles with the size of approximately 1 nm can aggregate almost without any activation energy, which causes the formation of aggregates.

In order to explore the properties of independent particles, their surface is passivated by forming complexes with various polymers. Figure 1.2 shows images of passivated gold nanoparticles received by electron microscope. The distribution of particles by size is also shown in Figure 1.2. The enveloping curve is a function

$$f(R) = 1/\sqrt{2\pi R\sigma}\ \exp\!\left[\ln\left(R - R_0\right)^2/2\sigma^2\right]$$

where R is the diameter of the particles. The most accurate match between the theoretical and the experimental results is achieved at the average particle diameter $R_0 = 1.9$ nm and the standard deviation $\sigma = 0.2$ nm.

Nanoparticles easily interact with other chemical compounds, which is followed by the formation of substances with new properties. Even Au_{55} is usually very reactive and has a very short lifetime in the free state. Aluminum nanoparticles also have a high chemical reactivity. If an isolated aluminum nanoparticle is placed in the air, it is immediately oxidized and covered with a layer of oxide. As a rule, free metal nanoparticles can only exist in vacuum without any interaction with a medium.

It can be added that nanoparticles created by any method are usually obtained in a metastable state. On the one hand, it makes the process of research and application of the particles for the creation of stable devices more complicated; on the other hand, the nonequilibrium of the system makes it possible to observe unusual and hardly predictable chemical transformations (see [9, introduction]).

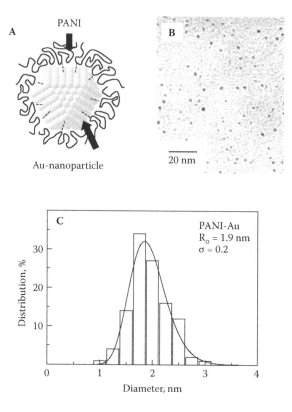

FIGURE 1.2 Gold nanoparticles stabilized with a polymer: (A) gold nanoparticles "protected" with hydrochloride polyallilamine (PAA); (B) the image of the particles (bar is shown in the figure); (C) distribution of the (Au-PAA) nanoparticles by size.

An important peculiarity of metal, alloy, and semiconductor nanoparticles is the catalysis of chemical reactions [10]. The high catalytic activity of nanoparticles is caused by electron and geometrical effects, though this division is conditional because both of the effects have the same source—the small size of the particles. The number of atoms in an isolated metal nanoparticle is small, so the distance between the levels $\delta \approx E_F/N$ (where E_F is the Fermi energy and N is the number of atoms in the particle) is comparable to the thermal energy of the particle $k_B T$. While $\delta > k_B T$, the levels are discrete, and the particle loses its metallic properties. The catalytic activity of the particles starts to appear when the δ-value is close to $k_B T$. In this case, the size of the particle at which the catalytic properties appear can be evaluated. For metals, the Fermi energy E_F is about 10 electron volts (eV) at a temperature of about 300 K. With $\delta \approx E_F/N = 0.025$ eV, then $N \approx 400$. A particle containing 400 atoms has a diameter of approximately 2 nm. Most of the data [10] testify that catalytic properties start to change as the size of particles reaches 2–8 nm. At the same time, because the electron structure of a nanoparticle depends on its size, the ability to enter into reactions with other substances also depends on the size [9].

Besides the electron effect described previously, there is another effect caused by the fact that, in small particles, the percentage of atoms placed on the surface that can have an electron configuration different from the interior atoms is large. This geometric effect also causes changes in the catalytic properties. The geometric effect depends on the ratio between the number of atoms placed on the surface, on the edges, and on the apexes of the particle that have different coordination. When atoms with small coordination are more catalytically active, the catalytic activity increases with a decrease in particle size. If the atoms with large coordination placed on the edges are more active, the increase of the rate of the catalyzed reaction will be observed with an increase in particle size. As a rule, nanoparticles show catalytic activity at a very narrow range of sizes; at the same time, the kinetics of the reaction in nanosized systems with limited geometry can be different from the typical kinetics (see [9, introduction]).

It is also important to point out that the carrier plays a certain role in the catalysis, because the atoms of the catalyst that enter into direct contact with the carrier can change their electron structure as a result of the formation of bonds to the carrier. On the whole, it can be noted that in the case of nanoparticles, one can carry out reactions that are not possible using the same substances in their usual condition.

The advanced surface of isolated nanoparticles significantly affects their properties. In the case of nanoparticles, it is necessary to consider the dependence of surface tension on particle size, because the surface energy is a significant quantity in comparison to the volume energy [10]. As particle size decreases, the contribution of the surface energy F_s to the free energy increases:

$$F = F_v + F_s$$

where F_v is the volume contribution.

If phase 1 is stable in bulk samples (as opposed to samples at the nanoscale) at a certain temperature, i.e., if the condition $F_v^{(1)} < F_v^{(2)}$ is observed, then with a decrease in particle size F_s it is possible that

$$F_v^{(2)} + F_s^{(2)} \leq F_v^{(1)} + F_s^{(1)} \tag{1.2}$$

and at a relatively small size, phase 2 will be stable.

It follows from Equation (1.2) that any deformation of the nanoparticle structure that reduces surface energy can decrease the total energy of the system. Such changes occur as a result of changes in the structural parameters of the packing of particles compared with the packing typical for the bulk material. While the structure of relatively large particles of a disperse phase is generally analogous to the structure of the corresponding phase, their structure can be different in the case of nanoparticles [5]. Because the surface energy is minimal for densely packed structures, the face-centered cubic or hexagonal structure is the most preferable for nanoparticles. At the same time, the transition from bulk materials to nanoparticles can be followed by changes in the distances between atoms and the periods of the lattice. The principal question is whether the periods of the lattice increase or decrease at the reduction of particle size and, if so, at what size this change becomes evident.

As noted previously, as particle size decreases, the contribution of surface atoms to the energy of the system increases, which leads to the appearance of a number of thermodynamic consequences, such as the dependence of the melting temperature of nanoparticles on their size. This effect can be described by the following equation:

$$(T_m - T_m^*)/T_m^* = \Delta T_m/T_m^* = -2V_m(L)\gamma_{sl}/\Delta H_m R \qquad (1.3)$$

known since 1871, when Thomson showed that the melting temperature changes inversely to the radius of the particles. This equation is known today as the Gibbs–Thomson equation, where T_m is the melting temperature at particle radius R; T_m^* is the melting temperature of the bulk material; $V_m(L)$ is the molar volume of the fluid; γ_{sl} is the surface tension between the solid and the liquid layers; and ΔH_m is the melting heat of the bulk material.

Equation (1.3) shows the decrease of melting temperature with decreasing size of nanoparticles. This behavior is illustrated in Figure 1.3, which contains data on gold nanoparticles (Figure 1.3(A)) and tin nanoparticles (Figure 1.3(B)) of different diameter. In particular, it can be noted that T_m of a massive gold sample decreases from 1360 K to 400 K in the case of gold nanoparticles with a diameter of 2 nm. A decrease of melting temperature of nanoparticles has been observed for such metals as Ag, Cu, Al, In, Ga, Sn, Bi, and Pb (see [11, introduction]). It is also known that in the case of CdS nanoparticles, the T_m decreases from 1650 to 400 K as particle size decreases.

According to Equation (1.3), the dependence of the melting temperature on $1/R$ for nanoparticles is a straight line, and the melting temperature of the bulk material, namely, gold can be obtained by extrapolating this line (Figure 1.4); the applicability

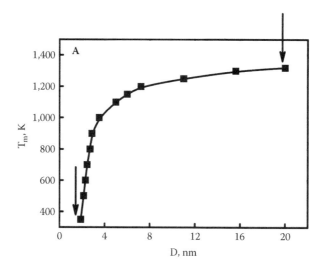

FIGURE 1.3 The dependence of the melting temperature (T_m) on the size of gold nanoparticles (A) and tin (B). A, the particles with different size are marked with the arrow for comparison.

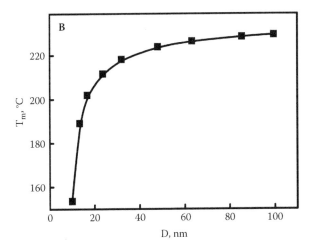

FIGURE 1.3 **(Continued)** The dependence of the melting temperature (T_m) on the size of gold nanoparticles (A) and tin (B). A, the particles with different size are marked with the arrow for comparison.

of the Gibbs–Thomson equation for estimating the melting of gold nanoparticles is illustrated in Figure 1.5.

Note that as nanoparticle size decreases, the term *phase* become less distinctive: The boundaries between the homogeneous and heterogeneous phases—between the amorphous and crystalline states of the substance—are difficult to determine. (The melting itself is a phase transition, i.e., a cooperative process.) At a small number of atoms in a nanoparticle, a phase transition is less distinctive: It stops being sharp, and as the number of atoms in the particle decreases, the melting curve itself starts changing its shape. Only at the thermodynamic limit of an infinite number of atoms does a sharp

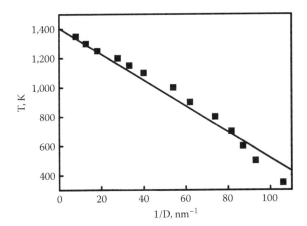

FIGURE 1.4 Dependence of the melting temperature (T_m) of gold nanoparticles on $1/D$ (D = diameter of gold nanoparticles).

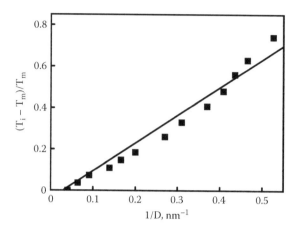

FIGURE 1.5   Dependence of $(T_i - T_m)/T_m$ on $1/D$ value.

phase transition—analogous to the phase transition in a bulk material and described by an S-shaped melting curve—cease to be observed in a nanoparticle. Consequently, there are quite simple and convenient experimental criteria that can be used to distinguish the properties of nanoparticles from the properties of bulk materials.

This means that, in contrast to bulk materials, the size of a nanoparticle can be considered an equivalent of the temperature that determines the behavior of the whole system. Thus, the surface properties of nanoparticles determine both their stability and their reactivity.

1.1.2   PLASMON RESONANCE AND THE "QUANTUM SIZE EFFECT" DETERMINE THE OPTICAL PROPERTIES OF NANOPARTICLES

As stated previously, nanoparticles have principally different physicochemical properties than larger structures, which can be explained first of all by the presence of the geometric size effect. However, the quantum-mechanic effects that start to appear in the behavior of nanometric objects are much more important [11].

The optical properties of metals and semiconductors are significantly different, as determined by the different locations of the conduction band, the valence band, and the Fermi level. Consequently, the size effects are different for metal and semiconductor nanoparticles.

The scattering and the absorption of the light in nanoparticles has a number of peculiarities in comparison to macroscopic materials. In particular, as metal nanoparticles absorb light, the absorption bands—absent for the bulk materials—start to appear in the visible region of the spectrum. For instance, films consisting of aluminum particles with a diameter of 4 nm have a distinctive absorption maximum at 560–600 nm [10]. Analogous maximums are typical for Ag, Cu, Mg, In, Li, Na, and K nanoparticles.

The optical properties of colloid solutions of metal nanoparticles are determined by the collective excitation of delocalized electrons. Under the effect of an electromagnetic field, the delocalized electrons (conduction electrons) in the nanoparticle are displaced against the positively charged frame. As a result of this displacement, a back-moving force inverse to the value of the displacement looks like that seen in a harmonic oscillator. As the frequency of the oscillation of electrons and the frequency of the exterior field coincide, a resonance effect caused by the excitation of the electron oscillations is observed. The description of the collective movement of electrons in terms of quantum mechanics leads to the term of elementary excitations, *plasmons*, which have a certain energy. The collective excitation of surface electrons is called "surface plasmons" (see [11, introduction]).

The plasmon effect, i.e., the resonance absorption of the incident electromagnetic radiation by nanoparticles, leads to the unique color gamut of colloid solutions of noble, alkali, and rare-earth metals. In particular, colloid solutions of gold nanoparticles can be orange, purple, red, or green, depending on the size of the particles. The plasmon effect in gold nanoparticles gives the beautiful red color to medieval stained-glass windows [11].

As nanoparticles have a large surface–volume ratio, the position of the plasmon resonance band is very sensitive to the properties of the medium boundary. Any changes in the environment of the particles, such as modification of the surface, a change in the dielectric constant of the medium, aggregation, etc., leads to changes in the color of the dispersion. Aggregation of nanoparticles causes consolidation of the plasmons and a significant displacement of their frequency. This causes a unique sensitivity at the surface of nanoparticles, which makes it possible to use metal nanoparticles as sensors.

The optical properties of the quantum dots (nanoclusters)—nanoparticles of semiconductors (for example, cadmium selenide) that behave as isolated atoms—are also interesting. Quantum dots contain several hundred atoms, and both the optical and electrical properties of these structures are different from the properties of macroscopic materials [12]. Quantum dots can absorb light waves, moving electrons to a higher level, and emit light as electrons move to lower levels.

It is known that in a bulk solid body with semiconducting properties, electrons can be located on levels that form bands. The excitation of electrons by light exposure can move an electron from a lower valence band through a band-gap and into the conduction band. In semiconductor nanoclusters with up to several atoms in a cluster, as well as for bulk materials, there is an energy gap between the valence band and the conduction band that determines the absorption and the emission of light. In this case, the light excitation is considered in the framework of the excitation of an exciton, which is a weakly bound electron-hole pair. At the same time, the electron appears in the conduction band, and the hole appears in the valence band. In the case of a nanocluster, the size of an exciton can be comparable or exceed the size of a cluster; in terms of the size of carriers (electrons and holes), this means a quantum limitation. With a decrease in the size of the cluster in comparison to the bulk material (see [11, introduction]), the width of the energy gap (the forbidden band) must increase, and the energy of the transition increases ($E \approx 1/R^2$, where R is the cluster radius), which causes the displacement of the bands in the spectra of absorption and luminescence corresponding to the electron excitation to the area of higher energies (blue displacement). At the same time, the force of the oscillators (the intensity of absorption) concentrates on several transitions. (The dependence of the spatial location of electron levels on the size of the cluster is called the "quantum size effect.")

It is known that every semiconductor has its own parameters of energy levels. However, they are different for nanomaterials and macroscopic materials. The smaller a quantum dot is, the broader is the gap between levels. For example, for gallium arsenide, the width of the forbidden band is 1.52 eV, but for a quantum dot of the same material containing 933 atoms, it increases to 2.8 eV, and for a dot containing 465 atoms it goes to 3.2 eV. Changing the size of quantum dots, one can change their color, even though the dots consist of one and the same gallium arsenide.

Figure 1.6 shows, as an example, a schematic view of a CdSe quantum dot. The diameter of such formations can vary from 2 to 10 nm. The CdSe "core" is surrounded by a ZnS shell, and the whole structure can be covered by a silicon oxide film. Some properties of the CdSe and CdTe quantum dots with different size and composition are compared in Figure 1.7 [10]. It can be seen that a decrease in the size of CdSe quantum dots is followed by a change in color from red to blue as the peak of the fluorescence band is displaced to the area of higher energies, i.e., a blue displacement is observed [9]. As the absorption boundary appears due to the presence of a gap, this means that the gap increases as the particle size decreases. At a first approximation, the energy of the band peak is usually inverse to the square radius of the CdSe particles. Moreover, it can be seen that, at a constant size of the quantum dots, a change in their composition is also followed by a change in their optical properties.

The recombination of charges generated by the light causes the luminescence of particles. The research on luminescence spectra of CdSe, CdS, ZnO, and ZnS nanoparticles has discovered a blue displacement, i.e., the displacement of spectra to the short-wave area with a reduction in the size of the particles [9].

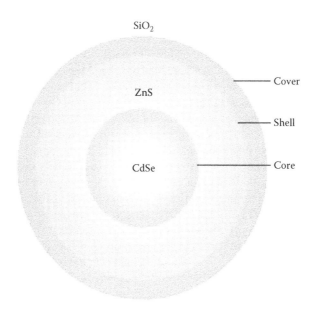

FIGURE 1.6   Schematic view of a quantum dot.

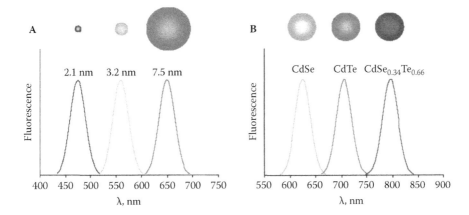

FIGURE 1.7 **(See color insert)** Some properties of quantum dots with various size and composition: three CdSe quantum dots with different diameter (A) and three quantum dots with different composition (average diameter is about 5 nm) (B) and their fluorescence spectra.

Thereby, nanoparticles of semiconducting materials are characterized by the presence of size effects expressed as the changes in their optical properties. Nanometric size and the transition from a massive material with zone structure to separate electron levels and the limitation of the free path of carriers by the effect of the surface of the nanoparticle (nanocluster) change the rules of selection and cause the appearance of new optical transition, as well as changes in the energy of transitions, the time of fluorescence and luminescence, and an increase of the force of oscillators (see [11, introduction]).

The nanometric size of a cluster also results in changes in electronic states. The transition from a massive material to a nanomaterial is followed by the division of electron areas (zones) into subareas (subzones) and separated electronic levels, while the transition from separate atoms and molecules leads to the broadening of separate levels to areas (zones), although these transitions are not very distinctive.

The appearance of discrete electronic levels is caused by the limitation of the free path of electrons and determines the quantum limitation effects in nanoclusters. This leads to a decrease in electroconductivity with a decrease in the size of a cluster and to the appearance of the Coulomb barrier for a one-electron transition between levels divided by the energy above kT (the product of the Boltzmann constant and temperature). The reduction in the size of a cluster is accompanied by the loss of metallic properties and the transition to a nonconductive state. Consequently, nanoparticles and nanomaterials based on them have a number of peculiarities that can be expressed in the presence of new optical and conductive properties.

In the case of nanoparticles, unlike bulk materials, there are multiple factors that mask the size effects. All of the peculiarities of nanoparticles mentioned previously can result in the appearance of specific points on the "scale relations" without monotonic changes in the properties of smaller particles. (Of course, it is also necessary to consider the distribution of nanoparticles by size.)

The size effect as a new parameter exerts a great influence on the development of the physics and chemistry of nanostructures [13]. The ascertainment of the size-effect mechanism is a topical problem for all types of nanomaterials. One can suppose an appearance of size effects, since the definite structure of nanoparticles defines a border between the description of nanosized objects in the framework of classical sciences and the description of nanoobjects in the framework of a new, real nanoscience.

Another problem related to nanoparticles is their stability, because thermal, chemical, deformative, and radiation exposure is often followed by complete or partial destruction of the unique properties of particles. This means that the reproducibility of the creation of nanostructures and their properties is an effect related to their stability to some extent. Therefore, the size of nanoparticles along with other thermodynamic parameters determines both the state and the reactivity of the system.

Summarizing all of the enumerated properties, it can be concluded that nanoparticles are characterized by new physical, physicochemical, etc., properties and functions that are significantly different from those of the initial structural elements and bulk materials [14]. Moreover, nanoparticles have unique and often unpredictable properties that open the possibilities for manipulations with those structures that cannot be performed under usual conditions. That is why it can be stated that nanoparticles involve more than size, because the dominating factor is not the size itself but, rather, the appearance of unique properties that are, in fact, functions of the size. One can add that these unique properties are immanent properties of all nanoobjects. Their absence means that new branches of modern physics must be developed to explain these cases.

It is clear from the data considered previously that the creation of nanoparticles and the research on their structures and properties, including the development of methods of manipulating these objects, as well as the application of nanoobjects in various systems, exceed the limits of a single science like physics, chemistry, or biology. Indeed, the science of the nanostructured state of a substance employs techniques and methods of multiple classical sciences, and very often this is the interdisciplinary convergence of basic sciences (such as chemistry, physics, biology, etc.) and applied sciences (such as material science and various areas of engineering).

Taking into account specific, size-dependent properties of nanoobjects, nanotechnology can be defined as an interdisciplinary science on the design and application of structured materials, devices, and systems whose properties are dependent on the nanoobject's geometry and the specific physicochemical features inherent to the nanoobject [15–18]. At the same time, "Nanotechnology is not nature. Nanotechnology is something that people create" [19]. The created nanomaterials can become integral parts of any macroscopic object.

1.2   BIOLOGICAL MOLECULES: A NEW BACKGROUND FOR THE CREATION OF NANOOBJECTS

Published results of recent international seminars and symposia (for example, see the proceedings of NanoTech 2005, May 8–12, Anaheim, California; papers from the NATO workshop "Nanomaterials for Application in Medicine and Biology," 4–5 October 2006, Bonn, Germany; the conference proceedings of "Nanobio and Other Perspective Biotechnologies," 15–18 October 2007, Puschino, Russia; and the proceedings of

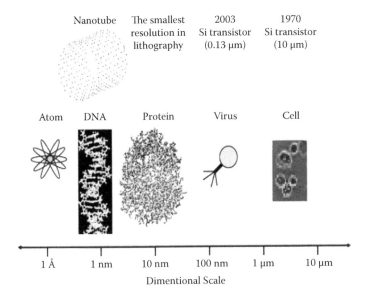

| Nanotube | The smallest resolution in lithography | 2003 Si transistor (0.13 μm) | 1970 Si transistor (10 μm) |

| Atom | DNA | Protein | Virus | Cell |

| 1 Å | 1 nm | 10 nm | 100 nm | 1 μm | 10 μm |

Dimentional Scale

FIGURE 1.8 The comparison of sizes of some technical devices and biological objects.

"Rusnanotech," 3–5 December 2008, Moscow) indicate that the number of nanotechnology studies using biological macromolecules has been increasing rapidly.

Considering the previously presented data describing the properties of various nanomaterials, it is interesting to compare the dimensions of biological molecules. From Figure 1.8, it is easy to notice that almost all of the biologically significant molecules are nanosized objects.

The interest in biological molecules is quite justified. During the process of evolution, these molecules have acquired a number of properties that make them extremely attractive for nanotechnology application. Before highlighting some of these properties, we first have to mention that the chemical diversity of building blocks, such as amino acids, lipids, and nucleotides (nucleosides), cannot be compared with that of inorganic components. Second, the biological building blocks themselves tend toward a spontaneous—and yet manageable, at the molecular level—formation of sophisticated spatial structures. Third, there are many ways of assembling (polymerizing) biological building blocks that make it possible to create a variety of nanostructures. The hierarchy of self-assembling biological structures begins with monomers (i.e., nucleotides and nucleosides, amino acids, lipids, etc.) that form biopolymers (such as DNA, RNA, proteins, and polysaccharides); their assemblies (membranes, organelles); and, finally, cells, organs, and even organisms. Fourth, nanobiomaterials that are formed as a result of self-assembly often have both upgraded characteristics and unique applications.

Given the combination of chemical reactivity of biopolymers and their ability to form hierarchical structures, the possibility of industrial production of biopolymeric molecules makes these molecules convenient for nanotechnology applications. Thus the use of biological molecules for the formation of artificial

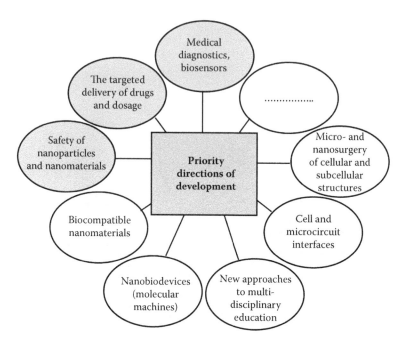

FIGURE 1.9 Prospective branches of nanotechnology that use biological molecules.

nanostructures, based on principles used by nature, seems quite obvious. It would be strange not to use the broad opportunities provided by nature itself for creation of biological nanostructures.

This has led to the appearance of new branches of nanotechnology, as shown in Figure 1.9. Note that there is an empty circle on the scheme, which indicates that the establishment of prospective branches of nanotechnology has not yet been completed. The scheme allows one to draw a number of important conclusions.

First, all of the branches employ biological structures (biological objects) to some extent. At the same time, two scenarios (A and B) of the development of these branches are possible.

Scenario A: The way from Nano to Bio involves the application of nanoparticles possessing specific size-dependent properties for research on biological objects and the eventual application of these nanoparticles in medicine and biotechnology. In this case, we are talking about the development of an area of science that can be called "nanobiotechnology." In both Russia and abroad, there are only a few examples of research in this area.

Scenario B: The way from Bio to Nano involves the creation of nanostructures using biological molecules (objects) that must receive specific properties before these nanostructures can be applied in medicine and biotechnology. In this case, we are talking about the development of an area of science that can be called "bionanotechnology."

Second, new areas of science that do not have a certain definition, namely, nano-biotechnology and bionanotechnology, are being established now. At the same time, it should be noted that, although many of the processes in living systems take place on the nanolevel, the consideration of all biological sciences, such as molecular biology, biochemistry, biophysics, immunology, virology, etc., as branches of nanobiotechnology remains controversial. That is why the discussions on the definition and the meaning of nanobiotechnology that started in 2000 are still ongoing [20]. To support this statement, several definitions of the new sciences by different authors are presented here. These definitions are far from perfect and are mostly relative:

2000. *Lowe, C. R.*: "Nanobiotechnology: the fabrication and application of chemical and biological nanostructures" [21].

2005. *Fortina, P.*: "Nanobiotechnology is the convergence of engineering and molecular biology that is leading to a new class of multifunctional devices and systems for biological and chemical analysis with better sensitivity and specificity and a higher rate of recognition" [22].

2006. *Medvedeva, N. V.*: "Nanobiotechnology (or biomolecular nanotechnology) is biological application or use of nanotechnology. The goal of nanobiotechnology is to understand the functioning principles of biological molecules (biological units) in order to create minor components of life with the use of special materials and interfaces" [23].

2006. A *Glossary of Biotechnology and Nanobiotechnology Terms* was published in 2006 [24]. A comment made by one of the reviewers of this reference book is most interesting and deserves to be quoted. Eleanor Randall wrote: "As I began this review, I needed a clearer definition of nanobiotechnology and its relationship with bionanotechnology. Interestingly, the title term nanobiotechnology was not included in the Glossary and, in the preface, one finds the term bionanotechnology used" (see http://www.istl.org/06-spring/review1.html and http://www.biomedical-engineering-online.com/content/6/1/5).

Indeed, page 45 of the glossary says: "Bionanotechnology: refers to the application of biotechnology within the fields of nanotechnology"

This quote shows that a well-known American glossary—one that is still relevant—gives no definition of *nanobiotechnology*; there is only a definition of *bionanotechnology*. The question whether the term *bionanotechnology* is a synonym for *nanobiotechnology* is still open, but a thorough analysis shows significant difference between these terms. For example, "Bionanotechnology can be called a nanotechnology (with its laws and rules) that uses biological objects as building blocks."

Third, Figure 1.9 demonstrates that the combination of nanotechnology and medicine is quite remarkable. The importance of this area of nanotechnology is substantiated not only by EuroNanoForum-2005, whose program was concentrated on the connection between nanotechnology and the health of European Union (EU) citizens, but also by the conference "Nanotechnology in Oncology" (6 December 2006, Moscow, Russia). Namely, Doctor Ottilia Saxl (Institute of Nanotechnology, UK), one of the founders of the EuroNanoForum, said: "Medicine is the area of nanotechnology that attracts attention of specialists from different areas. The idea that nanotechnology may make treatment of many diseases more targeted and purposeful corresponds to the interests of both doctors and their patients."

The idea of using nanotechnology to target the treatment of diseases seems most attractive. Indeed, the distribution of drugs (medicines) in the body of a patient can be regulated so that they can reach only a designated destination point. This can be achieved by applying nanostructured carriers, also called "nanoparticles for target delivery of drugs," which are created considering the biochemical status of an individual patient's body. Given that the unique properties of nanoparticles involve their highly developed surfaces (in comparison with those of traditional materials), nanoparticles make it possible to overcome the poor solubility and adsorptive properties of conventionally administered drugs. At the same time, nanoparticles are used to cure various pathologies as well as to diagnose diseases. The creation of new nanomaterials with upgraded therapeutic and diagnostic applications is an important goal.

Nanomedicine is the most rapidly developing area of research and applications in the field of nanotechnology. A great number of reports at the recent International Rusnanotech Forum (3–5 December 2008, Moscow) and the Nanotechnology in Oncology conference (6 December 2008, Moscow) were devoted to the use of nanotechnology approaches to solve medical problems.

As biological molecules continue to attract the attention of researchers as a potential basis for the creation of nanoobjects with new properties, the interesting question is: Should the dimensional properties of nanoparticles be considered if biological molecules—consisting of a large number of atoms of various chemical elements—are used as construction blocks?

Before answering this question, it is necessary to return to Figure 1.8. The comparison of sizes of biopolymeric molecules given in this figure allows one to make a number of remarks. First, all of the biopolymer molecules, such as nucleic acids or proteins as well as viral particles, have sizes that correspond to the nanoscale. An interesting remark is made in a work by Amato [19]: "All biological objects that have subcellular size are in nanometer range. A typical protein, like hemoglobin, has a diameter of about 5 nm; the diameter of a DNA double helix is close to 2 nm, and mitochondria have a size around a hundred nm." This suggests that the use of the nano- prefix in such terms as *nanoproteins, nanochitosan, nanoantibodies,* and *nanoviruses* makes no scientific sense. These terms are used, as a rule, only to attract the attention of a scientific audience to performed experiments in order to obtain further financial support, and nothing more.

Second, it follows from Figure 1.8 that the idea of an upper limit of nanosized biological objects becomes vague. The fact that the definition of various nanomaterials given by 7th International Nanotechnology Conference (2004, Wiesbaden) includes nanopolymers, biomaterials, different tubular objects, supramolecular structures, and catalysts testifies in favor of the vague definition of the size of biological nanoparticles [25]. The upper size limit of these structures is mostly relative, while the lower limit is determined by the size of molecules and atoms.

To answer the question regarding the dimensional properties of nanoparticles to be considered for biological molecules, it is interesting to observe the changes in

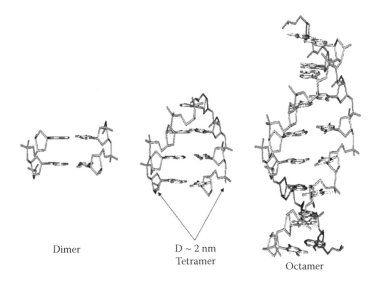

Dimer

D ~ 2 nm
Tetramer

Octamer

FIGURE 1.10 **(See color insert)** Hypothetical scheme of the formation of a helical structure from nucleotide molecules. (Kindly provided by S. Basili, Padova University, Italy.)

the properties of structures formed by nucleoside molecules—integral elements of nucleic acids (Figure 1.10). The figure on the left shows the structure of a hypothetical dimer molecule: One of the chains consists of a dinucleotide containing two adenosine (A) residues, i.e., it has the following structure: A-A. The other (complementary) chain consists of two thymine (T) residues, i.e., its structure is T-T. The probability of formation of such a structure under standard laboratory conditions is quite low, and the initial dinucleosides behave as isotropic molecules that can be described only by the distance between the adjacent nitrogen base pairs, which is 0.34 nm. If the initial chain of each nucleoside contains four base pairs, i.e., one chain has the (A-A-A-A) structure while the complementary chain has (T-T-T-T) structure, then the created structure begins to obtain the features of a double-stranded structure with a diameter of ≈2 nm (the structure in the center of the figure). When the initial chain of each of the nucleosides contains eight base pairs, i.e., one chain has (A-A-A-A-A-A-A-A) structure and the complementary chain has (T-T-T-T-T-T-T-T) structure, the created structure obtains the properties of a double-stranded helical structure similar to the structure of a natural nucleic acid (on the right). It can be said that as the number of nucleoside pairs increases, an ordered structure with parameters corresponding to the nanometric scale appears. The properties of this structure are significantly different from those of the initial elements.

The thermal destruction (melting) of spatial structures of double-stranded oligonucleotides created from the enumerated structural elements is even more interesting [26]. The melting curves of synthetic biopolymeric molecules containing different numbers of nucleotide pairs are compared in Figure 1.11. The melting temperature (T_m) of a very long double-stranded poly(A)×poly(U) molecule (a massive material) is 49°C. Here, the hyperchromism is expressed as $100(OT - OD)/OD$, where OT is

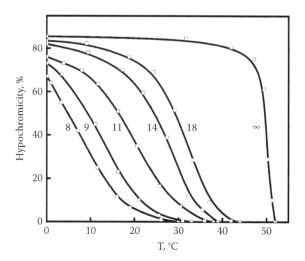

FIGURE 1.11   Changes in the shape of the melting curve and melting temperature of double-stranded oligo(A) × oligo(U) at the increase of the number of nitrogen base pairs (from 8 to ∞).

the optical density at a fixed temperature and OD is the optical density (at 260 nm) of denatured poly(A) × poly(U). The concentration of oligonucleotides was varied from 5 to 11 mM in 50 mM Na-cacodylate buffer at pH 6.9.

From Figure 1.11, it is easy to see that as the number of nucleotide pairs in the polymeric molecule increases, the shape of the melting curve of the forming helical oligonucleotides changes from exponential to S-shaped. Moreover, it is obvious that an increase in the number of nucleotide pairs increases the melting temperature, which reaches a maximum for very long polynucleotides (bulk materials). A similar increase in the T_m values to a maximum—corresponding to that of bulk materials with an increase in the number of building blocks in the chain—is typical as well for double-stranded DNA molecules [27].

In the forming structure shown in Figure 1.12, the dependence of the melting temperature on the number of nucleotide base pairs is similar in shape to the dependence of the melting temperature of gold or tin nanoparticles on their size (see Figure 1.3). At the same time, the relation of the melting temperature to the number of base pairs in the nucleotide chain, i.e., to the size of the structure, is described in the same way as for the metal nanoparticles in the framework of the Gibbs–Thomson equation (Figure 1.13).

Comparison of Figure 1.12 and the lines given in Figure 1.13 shows that in the case of a nanoobject, i.e., a double-stranded polynucleotide molecule assembled from biologically significant elements, the previously described properties typical of metal nanoparticles are retained.

The example given in Figure 1.11, known since 1971, shows that when nanosized structures are formed by biological molecules, some properties characteristic of a nanoobject (except the size) can be detected.

These results allow one to draw several conclusions:

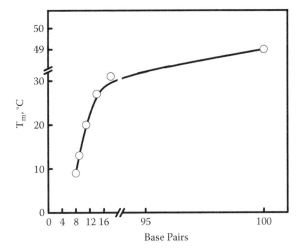

FIGURE 1.12 The dependence of the melting temperature of double-stranded oligo(A) × oligo(U) upon the number of base pairs.

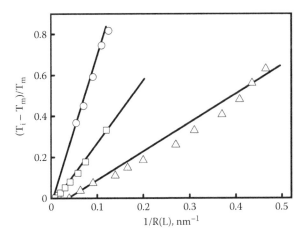

FIGURE 1.13 Dependence of $(T_i - T_m)/T_m$ upon $1/R(L)$ values for different nanostructures. L = the total length of base pairs in the oligonucleotide chain; 2 and 3 = the diameter of gold and tin nanoparticles, respectively.

1. The properties of nanoparticles with different sizes can be detected by various physicochemical methods.
2. The presence of nanoparticles can be detected by various physicochemical properties usually absent in the initial materials. These include:
   - The dependence of thermodynamic parameters (the temperature of phase transitions, the shape of the melting curve) on particle size of nanoparticles
   - The dependence of the chemical reactivity or the catalytic activity on particle size

- The dependence of electrochemical or magnetic properties on particle size
- The dependence of structural (crystallographic) parameters on particle size
- The dependence of quantum (optical) effects on particle size

3. Based on this list, it can be asserted that completely new effects, such as the ranging of the properties depending on the size of structures, the domination of surface properties, significant energy saturation that determines the high chemical reactivity of structures, quantum effects, etc., become apparent and prevail in nanoparticles that consist of a limited number of construction blocks.

4. It is also interesting to come back to the evaluation of the particle sizes at which the specific effects can be observed. The evaluations given in the work of Andrievski and Glezer [8] show that

the size effects in nanoparticles can be quantum by character, if the size of a particle is comparable to the de Broglie (λB) wavelength (λB = h/($2m^*E$)1/2), where: h - Planck constant, m^* - the efficient mass of an electron, E - energy of an electron. Using the known values of m^* and E, it can be proved that the quantum size effects will be observed in metals if the size of nanoparticles is below 1 nm. For semiconductors (especially with a narrow "forbidden" band, such as InSb) and "semimetals" (such as Bi) λB value is much higher and may reach about 100 nm.

The work of Pitkethly [4] contains a slightly different statement:

Materials with the size of particles from 1 to 250 nm are in the area where both quantum effects and properties of a massive substance are observed. Thereby, many physical properties of materials are controlled by effects that depend on the size in this area.

A larger quantity (below 500 nm) used by Russian Nanotechnology Corporation was given in the same report [4, p. 17].

The fact that the existence of scale effect is possible for objects formed from both atoms and molecules is testified by the definition for size effects given in the work of Pool and Owens (see [9, introduction]):

The size effects are the effects expressed in qualitative changes in physicochemical properties and reactivity depending on the number of atoms or molecules in a particle that take place in the range below 100 atom- or molecule-diameters.

Assuming (from the numbers provided here) that the size effects can be retained for particles with an average size of 250 nm, it would be interesting to compare this size to the length of a DNA molecule at which the molecule retains its rigid linear structure. It is known that the persistent length of a double-stranded DNA molecule that consists of ordered structural elements (nucleotides) with similar properties is close to 50 nm [27–29]. The comparison of these two numbers (50 nm and 250 nm) testifies in favor of the simple assumption that, in materials created from rigid linear DNA molecules and their complexes with various biologically active compounds that retain the monomeric order of nucleotides, the size effects can be observed.

5. The reason for the absence of size effects (specific properties) in other biological objects, even those that have a nanometric size, may be caused by both the globular spatial structure of these objects and the fact that the globule itself consists of a set of building elements (amino acids) with significantly different properties. It can be added that such biological objects can exist without the use of the *nano-* prefix, and most of them are subject to research within the framework of classical sciences: molecular biology, biophysics, virology, immunology, polymer chemistry, colloid chemistry, etc.

1.3 ORGANIZATION OF RESEARCH IN NANOTECHNOLOGY AREA

As the significance of nanotechnologies is interpreted to various strata of society, the recognition of the key role these technologies already play and will play in the future has led to the creation of national nanotechnology programs in every developed country [30]. These long-term programs are financed from both governmental and private sources. Such initiatives have resulted in intense competition between countries in the area of nanotechnology. Namely, the National Science and Technology Council of the President of the USA (founded in 1993) has formed the Interagency Working Group on NanoScience, NanoTechnics and Nanotechnology (IWGTN) that developed a program called the National Nanotechnology Initiative (NNI). The project was supported by President Clinton in 2000 and resulted in a rapid increase of investments in the development of nanotechnologies in the United States. The governmental financing of research conducted by various agencies within the framework of the NNI indicates the high volume of financing of works in the field of nanotechnology.

In an effort to popularize the NNI among various strata of society, soon after it was approved in 2000, a few founders of the NNI program—M. C. Roco, R. S. Williams, and P. Alivisatos—published a book that was translated into Russian and published in 2002, called *Nanotechnology Research Directions: Vision for Nanotechnology R&D in the Next Decade* [30]. In this book, the leading participants of the NNI program explained the importance of nanotechnology for various branches of industry and science in the United States. The book gives a clear conception of both the principal branches of research in nanotechnology and the major "characters" of the project. The total number of publications in popular periodicals, both positive and negative, devoted to nanotechnology increased from 190 in 1995 to 7,000 in 2003; according to a report by Lux Research Agency, the number reached 12,000 in 2004. The journalistic activity of various agencies in the area of nanotechnology has attracted the interest of the community and significant investments.

In expectation of a large market, the investment activity of corporations has increased. Lux Research supposes that, in the near future, most of the financing for the development of nanotechnologies will be provided by corporations. In North America, they already invest more in nanotechnology than various local and federal government offices—$1.6–1.7 billion. In Asia, the figure is $1.4–1.6 billion, and in Europe, the numbers are $650 million to $1.3 billion. Clearly, there is a flurry of investment activity in this sector. Shares of some companies go up only because they add the *nano-* prefix to their names, though most of the companies have hardly ever done any nanotechnology research. In 2004, 1,500 companies around the world announced the beginning of research in the area of nanotechnology; 1,200 companies are novice enterprises, with almost half of them based in the United States. Lux Research points out that this is a "soap bubble," an assumption based on the fact that the assets of the companies, including those invested by venture funds, are proportional to the results of their activities. It means that nanotechnology innovations that affect the scopes of many activities can significantly change the situation in the established production chains in the next few years. (The technical and engineering sectors have the greatest expectations in the application of nanotechnologies. Experts are absolutely sure that the further development of nanotechnology production will provide the possibility to create microchips with dozens of gigabytes of memory and processors with frequencies of several terahertz.)

The increased interest in nanotechnologies is not only typical for the United States. The research in this area is also broadly financed by the governments of Japan and the European Union (EU) countries. For example, in the spring of 2004, the European Commission promised to make a solid investment in the development of a new method of microchip production implementing the achievements of nanotechnology. The EU countries have chosen the path of development of scientific potential by integrating the efforts of the EU members. The 6th Framework Program, whose budget included 3.55 billion euros donated for nanotechnology and other related areas of science, became the mechanism of integration.

China is quickly becoming one of the leaders in the area of nanotechnology. The country has invested $300 million for the establishment of nanotechnology under the 5-year plan for 2001–2005. The number of nanotechnology-related patents in this country is increasing rapidly. In 2003, China was third in the number of such patents (after the United States and Japan).

According to forecasts of Lux Research, in 2010–2015, at least 2–3 million nanotechnology specialists will be needed to serve global needs for science and industry. As most of the leading universities take part in research in this area, it means that such specialists are being intensively prepared in developed countries. Special attention is paid not only to the training, but also to the organization of the cooperation between universities, private companies, and government laboratories; the importance of the interdisciplinary approach to the issues of nanotechnology is highlighted.

In many ways, the development of nanotechnology is very much like the information-technology boom that took place in the 1970s and 1980s or the biotechnology boom in the 1980s and 1990s. The rapidly increasing levels of investment show that the international nanotechnology train is gaining speed.

What is the situation like in Russia? The results of research performed by Russian scientists directly related to nanotechnology are in keeping with the international level and even excels in a number of areas. Speaking of the decisions taken at the governmental level, we should note the high interest and responsibility that various governmental offices have taken in the development of a national program on nanotechnology in Russia. The list of critical technologies (approved by the president of

Russia in March 2002) considers the application of nanosized objects and processes in some of the technologies.

Since 2003, it is possible to train specialists in some areas of nanotechnology, as the Russian Federation Ministry of Science and Education took the decision to create the nanotechnology sector as an experiment with two specialties: nanomaterials and nanoelectronics.

In October 2004, the hearings targeted at the development of principal areas of nanotechnology important for the country took place in the State Duma. As a result of the hearings, all of the participants, though coming from different agencies, expressed interest in the establishment of a federal program on nanotechnology.

On 5 May 2005, the presidium of the Russian Medical Academy determined in its resolution that nanobiotechnology applications in medicine are an important and promising area of research with significant scientific and practical value. The same resolution called for the creation of a nanomedicine program and the organization of special training.

In 2006, the Department of Scientific and Technical and innovational policy of the RF Ministry of Science and Education presented the project of a program on the development of nanomaterials and technology in Russia.

In 2007, the president of Russia, V. Putin, signed a decree establishing the Russian Nanotechnology Corporation. The program on nanotechnology covering activities of many of the Institutes of the Russian Academy of Sciences is under development. The yearly budget for nanotechnology research in Russia is almost equal to the American investment. All of these facts combined mean that 2007 has become the turning point in the establishment of nanotechnology research in Russia.

The informational support of various aspects of nanotechnology is performed by specialized periodicals, such as *Nano- and Microsystem Technics* (2004), *Russian Nanotechnologies* (2006), etc. Multiple publishing houses (such as Technosfera) publish monographs on various issues of nanotechnology. The nanoindustry concern has been established to conduct research in the area of nanotechnology. In 2004, a theoretical and practical conference "Nanotechnologies for Industry" took place. In 2008, the first international nanotechnology forum "RusNanotech" (Moscow) was held. A large conference, "Nanotechnologies in Oncology," takes place in Moscow annually.

These events show that the steps needed to establish a Russian national nanotechnology program are taken on different levels. They highlight the fact that only an active governmental policy in the area of nanotechnology—considering the fundamental knowledge in different areas of science as well as the interests of manufacturers and private investors—makes it possible to efficiently use the country's intellectual and scientific potential for developing new areas of science, creating new production facilities, increasing the level of health care, and ensuring the security of the Russian state.

At the same time, a number of circumstances deserve to be considered. Firstly, the question of the topics of the most urgent nanotechnology research is highly significant. The answer to this question determines the volume of investments in both fundamental research and practical development. The balance between these investments should be calculated very accurately, because any imbalance means that the implementation of a new nanotechnology product may become too costly. Secondly, regardless of the globalization, the development and the establishment of nanotechnology will be national by nature in any developed country. Considering the broad investments in nanotechnology and its importance for the economic development and defense capacity of a country, nanotechnology research will be at least partially closed. The created intellectual property will be secured by patents. This makes the idea that the knowledge on nanotechnology can be taken from open sources and applied for the development of one or another branch of nanotechnology in Russia inefficient. It means that Russian scientists will have to obtain fundamental knowledge independently, considering the volume of investments. Thirdly, interdepartmental barriers to the transfer of the accumulated experience, different stages of research, and an imbalance in the innovational infrastructure may affect the volume of investments and lead to the scattering of resources to multiple areas. Finally, the content and the goals of the Russian Nanotechnology National Program are still little known to the society at large. The question regarding the extent that the key role of nanotechnology is recognized by different strata of Russian society is a cause of concern for Russian scientists and manufacturers. This is an issue of national importance because it may determine both the possibility of supplementary financing of certain areas of nanotechnology from various sources, including nongovernmental sources, and the efficiency of the development of the country in the near future.

1.4   THE HEALTH RISK OF NANOMATERIALS

Nowadays, the major efforts of scientists and manufacturers, as well as major investments, are targeted toward the creation of nanomaterials and the development of nanotechnologies. The appearance of various nanomaterials has cast new colors on the rapid development of nanotechnology. By October 2008, the international register of nanomaterials (www.nanowerk.com) contained over 2,000 objects, and within the past year the number of objects has increased by about 60%. It is obvious that the number of new nanomaterials will only increase in the near future, which will lead to an increase of contacts between humans and nanomaterials as well as an increase in the amount of nanomaterials entering the environment.

As with any other new technology, nanotechnology represents both unquestionable advantages and a potential danger to human and natural ecosystems. There is an issue that may limit the broad application of nanomaterials: the toxicity of these materials. This problem is mostly based on the consideration of the experience of the use of asbestos fibers [31]. Indeed, the inhalation of asbestos fibers causes progressive lung disease, lung cancer, etc. At the same time, the oncogenicity of asbestos strongly depends on the size of the fibers, the ratio between fiber length and diameter, the spatial charge of fibers, etc. [32]. The risk related to the use of asbestos in the past makes us reevaluate the importance of the detection and calculation of potential toxicity of any created nanomaterials.

Currently, the toxic properties have been described for particles with a size from 0.1 to 1 nm (such as steams, gases, molecules suspended in a solvent) and aerosol particles greater than 1,000 nm. Nevertheless, even with sufficient knowledge of the toxicology of a certain material, its toxicity in nanosized form may be significantly different. The three factors, often acting simultaneously, that affect the toxicity of nanoparticles are [33]:

1. The shape of particles
2. The appearance of new functional groups on the surface and the chemical reactivity of the surface
3. The lifetime in a body, determined by poor solubility or slow removal

The combination of these factors indicates possible cytotoxicity of nanomaterials.

There is a principal recognition that nanoparticles of any materials are potentially toxic [34–36]. In this respect, two circumstances need to be considered. First, as discussed previously, the transition from macro- to nanosize causes changes in the properties of most of the materials, so nanomaterials may have completely different, often unpredictable physicochemical (and biological) properties than the initial substances. Second, nanomaterials may have unique bioavailability, making them potentially dangerous for humans. Unlike microscopic particles that can circulate in the air, nanoparticles can penetrate live organisms in different ways. Because of their size, nanoparticles can enter an organism through barriers impenetrable for larger particles [34–36]. This fact alone raises the issue of the potential harmful effects on human health that could be caused by the application of nanotechnologies and nanomaterials.

Currently, nanomaterials based on carbon are considered to be the key elements of nanotechnology, and these are attracting particular interest. Such materials exist in different forms, such as carbon nanotubes, carbon nanoparticles, fullerenes, nanodiamonds, etc.

Carbon nanotubes (CNT) are a new form of crystalline carbon (see Figure 1.14). Carbon nanotubes can exist as single-wall CNT (SWCNT) or multiwall CNT (MWCNT) structures that attract great interest due to their unique properties, such as mechanical durability, chemical stability, and high energy and heat conduction. These properties determine their applications in industry, electronics, and medicine. The world market for CNT reached $12 million in 2002 and increased

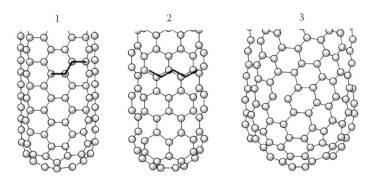

FIGURE 1.14 Computer model of single-wall carbon nanotubes with different structure: 1, "chair"; 2, "zig-zag"; 3, "spiral."

to \$700 million in 2005 (see *Carbon Nanotubes—Worldwide Status and Outlook: Applications, Applied Industries, Production, R&D and Commercial Implications*).

There are many examples of industrial application of CNTs, including their recent use as needles for atomic force microscopy, for targeted delivery of drugs, and for the specific adsorption of certain chemical compounds. CNTs have also been added to enhance polymers and other matrixes. Although biomedical applications of CNT are still at the research stage, it is known that they can be used in the creation of sensors, in detecting DNA or proteins, and as ionic channel blockers. It has been shown recently that SWCNT with a surface functionalized by amine groups can carry plasmid DNA, thereby increasing the expression of genes. These results mark the way for the use of CNT as a possible system of delivery of therapeutic and diagnostic molecules. At the same time, for a successful application as a drug delivery system, the length of CNT must be below 1 μm to penetrate cells efficiently.

However, there is a lack of information about the effects of carbon nanotubes and other nanomaterials on human health and the environment. This is due to the fact that CNTs are long thin structures with a diameter of several nanometers (though their length may reach several micrometers) that must have unusual toxicological properties reflecting the morphological peculiarities of both fibers and nanoparticles. In Figure 1.15:

A1= SWCNT aggregates

B1 = MWCNT aggregates

A2 = crude SWCNT bunches with included metallic nanoparticles

B2 = individual MWCNT

A3 = section of an SWCNT bunch containing more than 25 nanotubes

B3 = longitudinal section of MWCNT with empty central cavity and 20 walls
       on each side

The scientific community, on the whole, agrees about the toxicity of all carbon nanotubes because of their structural similarity to asbestos fibers [37, 38]. Recent research has shown that CNTs can aggregate in a stream of air, gas, or fuel, which means that humans may be exposed to daily effects of CNTs both indoors and outside.

CNTs are poorly soluble in water solutions, which limits their practical applications. One of the most widely known strategies used to increase the solubility of CNTs in water solutions is the introduction of reactive groups on their surface, which opens the possibility of biomedical applications of CNTs [39].

The question of how the parameters of CNTs (length, diameter, wall structure, surface charge, etc.) affect their toxicity remains poorly studied, although it is generally supposed that all of the parameters play an important role. To explore the effect of the shape of CNTs on the toxicity to cells, different types of cells were processed with CNTs with variable ratio of length to diameter (axial ratio), namely, MWCNTs with an average diameter of 20 nm and axial ratio 80–90, CNTs with an average diameter of 150 nm and axial ratio 30–40, and, finally, carbon nanoparticles (carbon black) with axial ratio about 1. All of the CNTs were suspended in a strongly diluted gelatin solution to avoid their aggregation. The experiments have shown (Figure 1.16) that all CNTs cause cell death [32].

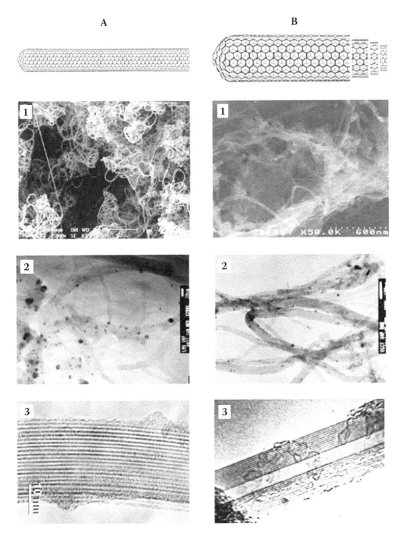

FIGURE 1.15 Single-wall (A) and multiwall (B) carbon nanotubes: A1, A2, A3 = SWCNT; B1, B2, B3 = MWCNT (images taken by electron microscope).

The typical image of H596 cells is given in Figure 1.16(A) as a control sample. The cells were dyed with hematoxylin and eosin: The nuclei look like purple inhomogeneous formations, and the cytoplasm is pale pink. Cell clusters are characterized by close cell-to-cell contacts, and individual cells have a polyhedral shape. The H596 cells after incubation with 0.02 fg/mL of MWCNT within one day are shown in Figure 1.16(B). The cells have lost their adhesion ability; the volume of the cells has decreased (arrows); and the nuclei have become smaller and denser.

Experiments evaluating excretion of SWCNT preparations (diameter 0.9–1.5 nm, lengths 0.5–3.0 μm) from kidneys after intravenous and endotracheal introduction have shown that, in the first case, after introduction of 0.35 mg/kg of ^{3}H-SWCNT,

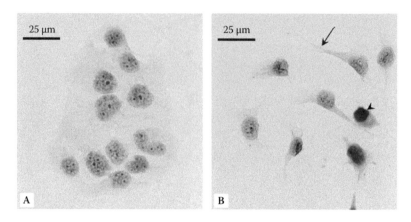

FIGURE 1.16   (See color insert) Cytopathologic analysis of human tumor cells (line H596).

radioactive excretions in the urine were observed within 30 days; starting with the second week, the level of the excreted radioactivity remained constant. The total monthly kidney excretion reached 9.5%–10.0% of the introduced dose of ^{3}H-SWCNT [40].

The effect of carbon nanotubes on human skin cells leads to stress and the accumulation of oxidation products [41]. Biochemical processes are accompanied by morphological changes in the cells (Figure 1.17).

It has been shown that CNTs functionalized with proteins are able to penetrate through plasma membranes of mammals and reach the cell nuclei. These CNTs can cause antigen-specific neutralization of reactions in vivo with the participation of antibodies [41].

CNTs become potentially dangerous for their creators and users because they penetrate the respiratory tract, they easily deposit in the lungs, and they relocate from the place of the initial landing and modify the structures of proteins [38]. In such cases, CNTs activate immunologic reactions that affect the normal functioning of organs.

Unfortunately, the information on the potential danger caused by the introduction of CNTs into animals is insufficient and remains a controversial subject. The establishment of in vivo pharmacological profiles for CNTs remains an important issue in the discussions about the safety of new nanomaterials.

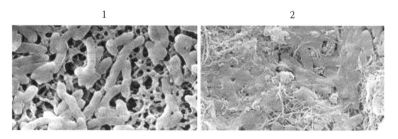

FIGURE 1.17   The effect of carbon nanotubes on *E. coli* cells during one hour: 1, in the absence of nanotubes; 2, in the presence of nanotubes (microscopic images).

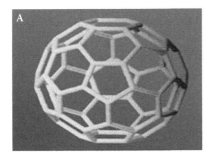

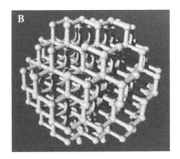

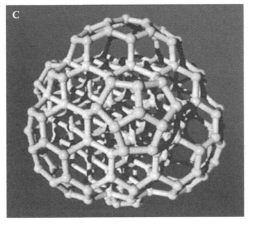

FIGURE 1.18 Computer models of nanodiamonds and fullerenes: (A) the image of a ^{60}C fullerene; (B) the image of a nanodiamond (1.4 nm); (C) the image of a cluster of 275 atoms (1.4 nm) with features of both a nanodiamond and a fullerene.

Regarding the other nanosized forms of carbon—fullerenes (Figure 1.18) and nanodiamonds (Figure 1.19)—there are several points worth noting. Fullerenes are unique polyatomic molecules with a diameter of 0.7 nm containing 60 or more symmetrically located atoms of carbon, with a polyhedral shape. Fullerenes were discovered in 1985 [42]. Initially, the structure containing 60 carbon atoms was called a *buckminsterfullerene* in honor of the American engineer and architect Richard Buckminster Fuller (1895–1983), whose constructions were based on the same principle. Scientists quickly contracted this to *buckyball*, because the spherical structure of C-60 looks like a soccer ball. C-60 has unique physical, chemical, electrical, and optical properties that can be changed by functionalizing carbon atoms. Fullerenes have no free valences, and even in a crystal, molecules are not bound covalently, which makes them strikingly different from any other known substance. A hydrophobic sphere, a large surface, and a high electron affinity make these substances very attractive as efficient antioxidants, carriers for targeted delivery of drugs and vaccines, and a means to give new properties to various medicines.

Many derivatives of ^{60}C fullerene have been synthesized, including some water-soluble substances (Figure 1.20). The methods of obtaining water suspensions of such substances have been developed, and the biological activity of many of them

FIGURE 1.19 Nanodiamonds (image taken by electron microscope).

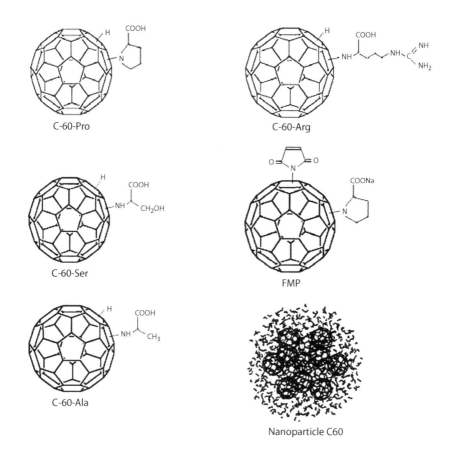

FIGURE 1.20 Water-soluble amino acid derivates of ^{60}C fullerene and a ^{60}C nanoparticle.

has been evaluated. It has been shown that they may have antitumoral, antiviral, antibacterial, and other activities. The treatment of macrophage cells with ^{60}C and ^{70}C fullerenes leads to the accumulation of tumor necrosis factors (TNF), interleukin-6 (IL-6), and interleukin-8 (IL-8) in cell supernatants [41].

Because the structure of fullerenes is different from structures of all the other known organic compounds, one of the questions of interest is: "Can the immune system recognize the carbon frame?" It is obvious that upon the introduction of fullerenes into the human body, the immune system may respond to fullerenes or to products of their interaction with proteins or other molecules [43]. As a powerful antioxidant, a fullerene can take part in cellular processes that involve the participation of active forms of oxygen, for example, neurodegenerative diseases, ischemia, allergic and cytotoxic reactions, and apoptosis [44]. Indeed, the preliminary results of research on the fullerene derivatives suggest the possibility of an allergic reaction [41].

The data presented here testifies that carbon nanomaterials are the materials of the future; hence, the practical application of these materials requires a thorough evaluation of their toxicity and the establishment of standards to avoid disorders in human health in the future. This also means that the principal mechanisms determining the toxicity of nanomaterials created from various forms of carbon nanoparticles need to be systematically explored.

Unfortunately, evaluation of the potential dangers of nanomaterials based on other nanoparticles has received little attention. At the same time, the need to evaluate the genotoxicity of all nanoparticles and nanomaterials is generally recognized. The probable genotoxic activity of nanomaterials is rooted in several causes, namely their ability to penetrate into living cells and induce free radicals of oxygen and nitrogen [45], to damage the cytoskeleton [46], and to reach the nuclei [47] and interact with DNA molecules [48]. The composition of some nanoclusters also contain atoms of compounds that have a carcinogenic effect, such as cadmium or arsenic salts [49]. Finally, the structures of some nanomaterials are similar to asbestos fibers [36], which have well-known genotoxic and carcinogenic effects. In addition, a comparison of the genotoxic effects of nano- and microparticles of the same compounds verifies that nanoparticles have higher activity [42]. The genotoxic properties of some nanoparticles have been studied in vitro.

Nanoparticles of realgar have been shown to increase the fragmentation of DNA and to induce apoptosis in U937 cells [50]. Chitosan molecules (chitosan nanoparticles) have induced DNA fragmentation and apoptosis in MGC803 human stomach carcinoma cells [51]. The genotoxic activity of nanoparticles is supposedly determined by their ability to induce radicals of nitrogen and oxygen that damage DNA [52], as well as high permeability and/or direct effects on intracellular structures, including the cytoskeleton and chromatin [53]. Though the genotoxic activity has been explored in some studies in vitro, the results of these works testify that nanoparticles may induce damage to the DNA structure, chromosome aberrations, formation of micronuclei, etc. Nanoparticles can be bound to biological macromolecules (DNA, RNA, proteins) and modify their properties, which means that the probability of genotoxic and mutagenic effects of nanoparticles and nanomaterials on a human organism can be quite high; such effects can lead to serious consequences.

Though the mechanisms of the action of metal nanoparticles on the human brain under normal or pathological conditions are unknown, there are data that point out the neurotoxicity of some particles. Namely, metal nanoparticles (Cu, Ag, or Al with a diameter from 50 to 60 nm) can cause a dysfunction of the human brain (decrease in cognitive abilities) as well as changes in the brain cells of healthy animals and aggravate brain pathology [54].

Research on the effect of TiO_2 nanoparticles on human lymphoblast [55, 56] and on bronchial epitheliocytes [57] testifies to the damage done to DNA molecules. It has also been shown that TiO_2 nanoparticles cause fibrosis and lung cancer in rats [55]. The results are most significant for producers of drugs and toothpastes, because TiO_2 particles are often added to such products as fillers, while the experimental results discussed here prove that the cytotoxicity of TiO_2 nanoparticles increases with decreasing size of the particles [58]. Experiments on the effect of cobalt-chromium alloy nanoparticles on human fibroblasts also testify to the damage to DNA molecules [45].

Due to their increased adsorption capacity, nanoparticles can act as carriers for some genotoxicants (heavy metals, pesticides, etc.) as they penetrate an organism. Nanoparticles with a high catalytic activity can initiate processes of free radical peroxide oxidation that cause damage to biological structures.

Nanoparticles can penetrate the human body in all the ways known: through the respiratory tract, the gastrointestinal tract, undamaged integument, olfactory nerve receptors, etc. At the same time, it is known that the pathogenic effect of inhaled particles depends on their shape [59]. The lungs, liver, gastrointestinal tract, brain, and kidneys are targets for nanoparticles. Nanoparticles accumulated in various tissues may cause chronic lung diseases, malignant neoplasm, malfunctions of cerebral and coronary blood circulation, heart malfunctions, etc. [60, 61].

The probable dangers associated with nanoparticles and nanomaterials is not in doubt. Professor Harry Kroto, who received the Nobel Prize in 1986 for discovering [60]C fullerene, says: "We must admit that mistakes and risks related to nanotechnologies are possible, though these are products of the 21st-century science."

It can be concluded from the results described here that nanoparticles have higher toxicity in comparison with usual macro- and microparticles, as they are able to penetrate cellular barriers as well as through the blood-brain barrier to the central nervous system. They can circulate and accumulate in organs and tissues, causing pathomorphological lesions of viscera (formation of lung granulomas, hepatic cirrhosis, glomerulonephritis), and nanoparticles are slowly excreted from a body [60–62]. Therefore, conducting research to determine the safety limits in the production and applications of nanomaterials, as well as developing documents regulating the conditions for working with nanomaterials, are very important practical issues of nanosafety.

It is obvious that nanosafety is a global problem. The research in this area has been conducted since the year 2000 in the United States (under the aegis of the Food and Drug Administration [FDA]), in the European Union (Organisation for Economic Co-operation and Development [OECD] and others), and in a number of international organizations (WHO, FAO, etc.). For example, in the United States, a special section was formed for the safety of nanomaterials (EHS, Health and Safety) within the framework of the NNI, which is a long-term program of scientific research in the area of nanotechnology. An infrastructure was developed for the analysis of this issue.

In 2004, the London Royal Society and the Royal Engineering Academy formed a workgroup to evaluate the conditions of nanoscience and nanotechnology and their place in modern society. The group has written a report concerning the potential dangers of nanobiomaterials. The report recommends care against inhaling a large number of nanoparticles (the employees of research laboratories may be exposed to this danger) and the emission of nanoparticles (especially the free particles not attached to a solid surface) into the environment. Currently, European standards automatically allow the usage of nanoparticles of a substance if the substance by itself is not forbidden. Nevertheless, considering the high chemical activity of nanoparticles, a substance that is not considered to be toxic may be dangerous as a nanopowder. Therefore, such materials need to be thoroughly explored to develop special standards for them. On the whole, the experts advise treating the application of nanoparticles carefully, verifying their safety repeatedly, conducting scientific assessment of commercial products, informing consumers, etc.

In 2004, The Policy Office of EPA (USA) formed an expert group to prepare the *White Book* (published in 2005) deliberating upon the potential risks of application of nanotechnologies. The experts emphasize that, although nanoparticles can accumulate in the air, soil, and water, there is no precise modeling of these processes. As there is little information about the consequences of uncontrolled emission of nanoparticles into the environment, the authors point out the necessity to fill this gap. Nanoparticles can be destroyed under the effect of light and chemical substances or as a result of contact with microorganisms, but these processes have been poorly examined. Nanomaterials easily enter into chemical transformations and form compounds with unknown properties, which means that special attention needs to be paid to risks related to nanoparticles.

It should be noted that, despite the broad support of nanotechnology development programs in developed countries, the administration of the Greenpeace organization calls for the public and the governments of the developed countries to declare a moratorium on further nanotechnology research until the toxicology research results are received.

Moreover, in 2007, the British Soil Association (SA), which certifies organic products, refused to give certificates for products containing nanoparticles, considering them to be dangerous. (SA is a nongovernmental organization founded in 1946 that currently has a charitable institution status.) SA cares for environmental safety and human health, supporting the use of harmless organic products. Namely, SA is involved in standardization and certification, and its standards are much more severe than the standards of the UK and the EU. The SA bans many of the genetically modified products. Of course, the ban does not have the consequences that a violation of the governmental standards would have, but the product does not get the "Organic product. Approved by SA" label. On 15 January 2007, the SA announced it would not certify cosmetics, medicines, food, clothes, and other products containing synthetic particles with a diameter below 125 nm. This action is explained by a concern for human health: If it is not known whether the danger exists, it is better to assume that it does. (SA points out that it was the first organization to ban nanoparticles because of its concern for consumer safety.)

The international discussions on the issue of nanosafety take place both abroad (for example, in the United States in December 2005 [63] and in Germany in April 2007 [64]) and in Russia, for instance, in the Institute of Crystallography of the Russian Academy of Science (RAS) in October 2006. This problem was a topic of particular interest at the International Nanotechnology Forum held in Moscow in 2008.

The participants at all of these discussions paid attention to the fact that the specific properties of any kind of nanoparticles (the advanced surface, high chemical reactivity and catalytic activity, the ability to travel easily in a stream of air or fluid, the absence of metabolism, etc.) may lead to serious consequences if the particles enter the environment or the body of an animal or a human. This demonstrates that the discussion of the rules of work with nanoparticles and the rules of their application in various areas of medicine, science, and technology has started [63–65].

The experts recommend conducting wide-ranging research on the potential risks and danger of the pollution of the environment with nanoparticles. It is necessary to find out how the biodegradation of nanoparticles occurs and how it affects the ecological chains in nature. The penetration of nanoparticles into the environment may have consequences that are hard to predict at this moment.

In Russia, nanosafety research has been conducted at the initiative of the Federal Service on Consumer's Rights Protection and Human Well-Being Surveillance since the end of 2006. The initial assumption is that nanomaterials should in any case be considered as new kinds of materials and products, and the description of their potential danger for the environment and human health is necessary according to the federal laws: "On Sanitary-Hygienic Well-Being of Citizens" and "On Quality and Safety of Products." Various institutes of the RAS, the Russian Academy of Medical Sciences (RAMS) and the Russian Academy of Agricultural Sciences (RAAS), as well as Lomonosov Moscow State University have developed a project—"Concept of toxicological research, methodology of risk assessment, methods of identification and quantitative definition of nanomaterials"—that was approved in 2007. To implement the provisions of the concept, a methodological instruction, "Safety assessment of nanomaterials," was developed for application in sanitary-hygienic supervision offices to assess the safety of nanomaterials and nanotechnology products for human health. The instruction entered into force in 2007.

In addition, to efficiently solve the problems enumerated here, the concept of medical-sanitary provision of nanoindustry work in the Russian Federation was developed in 2007 [66]. The concept is targeted at establishing a methodology of risk assessment for workplaces in the field of nanotechnology, for methods of identification and quantitative definition of nanomaterials and nanoparticles, and for specialized methods that would make it possible to evaluate the toxicity of nanomaterials and their influence on the environment and human health. The concept also provides for the development of hygienic safety standards and methods of sanitary supervision, as well as monitoring the health condition of personnel at production facilities where nanoparticles and nanomaterials are applied. The practical implementation of this concept will help to protect the health of the citizens occupied in the development, production, and application of nanoproducts. This confirms that the state of research in the area of nanosafety in Russia corresponds to the level in the other developed countries.

A number of observations about nanoparticles merit further discussion:

1. There is no doubt that some nanoparticles can exert a toxic effect on cells of various tissues. Multiple studies verify that the toxicity of a number of substances in nanoform can exceed the toxicity of the analogous solid phases or macroscopic dispersions. This is caused by physicochemical properties of nanomaterials in a superfine form, such as increased reactivity or catalytic activity, solubility, etc., as well as the penetration power of these forms.

2. The important issue is the diversity of nanoparticles, especially considering the fact that different forms of the same substance can exist. Many nanomaterials are produced by different technologies, which may result in differences in the effects of the created particles in the environment and on human health; this circumstance increases the risks that consumers of nanomaterials may face. The research on biological and toxicological effects of nanoparticles—depending on their size; the initial material, surface, charge, and other physicochemical properties; as well as the dose, ways of injection, concentration in the target organ, and the duration of exposure—is a complicated task. It is desirable that such research be conducted on living organisms, and not exclusively on cell cultures. Because the human population is genetically dissimilar, a large group of people may be extremely sensitive to the effects of nanoparticles. This emphasizes the need for special research to evaluate every form of a substance at a certain size for danger to human health. This will entail significant difficulties in evaluating the safety of practically important nanomaterials.

3. The development of uniform methodologies and research standards for evaluating the safety of nanomaterials is just beginning. The conventional methods used to evaluate the safety of medications may be ineffective in the case of nanoparticles. For example, using only special techniques, the scientists from the RAS Pharmacology Institute managed to detect undesirable genetic effects caused by zeolite nanoparticles.

4. The development of methods to evaluate the toxic effects of nanostructures and the need to conduct the requisite experiments will likely require both a substantial amount of time and significant funds. According to the calculations (2007) made by experts in the Project on Emerging Nanotechnologies organization, the total amount of money allocated for evaluating the danger of nanomaterials application in the United States was $39 million, which was only 4% of the total allocations for nanotechnology made by the U.S. Department of the Treasury. During the hearings before the U.S. Committee on Science in 2007, it was claimed that the costs for exploring the environmental and medical aspects of nanomaterials application must reach 10%–20% of the total governmental costs for nanotechnology. In this respect, it would be reasonable to establish scientifically grounded priorities [63–65] to predict the potential dangers of created and actually applied nanomaterials for human health based on the available information.

5. The social aspects of the issue, namely, establishing a sensible public perception of the dangers of nanomaterials, form a special category. An infrastructure for conducting research in the area of nanotechnology safety has already been formed in the developed countries. Indeed, a new branch of science—nanotoxicology—has been formed, with its own specialized publications such as journals, periodicals, textbooks, etc. Russia has only taken the first steps in this direction [67, 68].

There is a clear need to establish a complex research program targeted at ensuring the safety of nanotechnologies and nanomaterials for human health and the environment. It should be noted that, despite the evident limitations, nanotechnologies must ensure the high quality of life of the population and correspond to the social requests of the Russian Federation.

The measures to establish a national nanotechnology program in Russia are being taken on different levels. However, an active governmental policy in this area—based on the consideration of the fundamental knowledge of scientists with different specialties as well as the interests of manufacturers and potential investors—is needed to efficiently use the intellectual, scientific, and technical capacities of the country in the development of this new area of science and the creation of new industries based on nanotechnology.

1.5 SUMMARY

Considering a reasonably cautious approach to the potential dangers that evolve when creating and using nanoparticles of any composition, nanotechnologies provide a high potential for economic growth, increased quality of life, and conservation of energy and other resources. In short, development of nanotechnology corresponds to the social and economic needs of any society.

The achievements in chemical synthesis and biotechnology provide a fantastic opportunity for the creation of biomaterials and nanostructures that do not exist in nature and, moreover, to apply the principles and techniques used in nature for the creation of practical and important nanoobjects.

It can be expected that, as nanotechnology develops, biological molecules will be transferred from the world of biology into the world of technology.

REFERENCES

1. Birringer, R., H. Gleiter, and H.-P. Klein. 1984. Nanocrystalline materials. *Phys. Lett. B* 102:365–69.
2. Birringer, R., U. Herr, and H. Gleiter. 1986. Nanocrystalline materials: A first report. *Trans. Japan. Inst. Met. Suppl.* 17:43–52.
3. Andrievski, R. A. 2005. Size effects in nanomaterials: Rules and exceptions. In *Proc. 1st Iran-Russia Joint Seminar and Workshop on Nanotechnology*, 1–3. Tehran, Iran: Aftab Graphic.
4. Pitkethly, M. J. 2003. Nanoparticles as building blocks? *Nano Today* 6 (suppl. 1): 35.

5. Summ, B. D., and N. I. Ivanova. 2001. Colloid-chemical aspects of nanochemistry: From Faraday to Prigozhin. *Herald of Moscow State University, Chemical Bulletin* (Russian ed.) 42:300–5.

6. Murphy, C. J., A. M. Gole, J. W. Stone, P. N. Sisco, A. M. Alkilany, E. C. Goldsmith, and S. C. Baxter. 2008. Gold nanoparticles in biology: Beyond toxicity to cellular imaging. *Acc. Chem. Res.* 41:1721–30.

7. Andrievski, R. A., and A. M. Glezer. 1999. Size effects in nanocrystalline materials. I: Structure characteristics, thermodynamics, phase equilibria and transport phenomena. *Phys. Met. Metallogr.* 88:45–66.

8. Andrievski, R. A., and A. M. Glezer. 2001. Size effects in properties of nanomaterials. *Scr. Mater.* 44:1621–24.

9. Pool, Ch. P., and F. J. Owens. 2003. *Introduction to Nanotechnology*. New York: John Wiley & Sons.

10. Gusev, A. I. 2005. *Nanomaterials, Nanostructures and Nanotechnologies* (Russian ed.). Moscow: Fizmatlit.

11. Altman, Yu. 2006. *Military Nanotechnology* (Russian ed.). Moscow: Technosphere.

12. Maltsev, P. P., ed. 2006. *Nanomaterials, nanotechnologies, nanosystem technique: World achievements in 2005* (Russian ed.). Moscow: Technosphere.

13. Roduner, E. 2006. Size matters: Why nanomaterials are different. *Chem. Soc. Rev.* 35:583–92.

14. Yamamoto, Y., T. Miura, T. Teranishi, et al. 2004. Direct observation of ferromagnetic spin polarization in gold nanoparticles. *Phys. Rev. Lett.* 93:116801-1–4.

15. Yevdokimov, Yu. M., S. G. Skuridin, Yu. D. Nechipurenko, et al. 2005. Nanoconstructions based on double-stranded nucleic acids. *Int. J. Biol. Macromol.* 36:103–15.

16. Yevdokimov, Yu. M., and V. V. Sytchev. 2007. Nanotechnology and nucleic acids. *Open Nanosci. J.* 1:19–31.

17. Yevdokimov, Yu. M., and V. V. Sytchev. 2007. Nanotechnology and nucleic acids. *Technologies of Living Systems* (Russian ed.) 4:3–24.

18. Yevdokimov, Yu. M., and V. V. Sytchev. 2008. Principles for creating nanoconstructions with nucleic acid molecules as building blocks. *Uspekhi Khimii* (Russian ed.) 77:194–206.

19. Amato, I. 2005. Nanotechnologists seek biological niches. *Cell* 123:967–70.

20. Vogel, V., and J. Schloss. 2003. Nanobiotechnology. Report of the National Nanotechnology Initiative Workshop, 9–11 October, Arlington, VA.

21. Lowe, C. R. 2000. Nanobiotechnology: The fabrication and application of chemical and biological nanostructures. *Curr. Opin. Struct. Biol.* 10:428–34.

22. Fortina, P., L. J. Kricka, S. Surrey, and P. Grodzinski. 2005. Nanobiotechnology: The promise and reality of new approaches to molecular recognition. *Trends Biotechnol.* 23:168–73.

23. Medvedeva, N. V., O. M. Ipatova, Yu. D. Ivanov, A. I. Drozhzhin, and A. I. Archakov. 2006. Nanobiotechnology and nanomedicine. *Biomedical Chemistry* (Russian ed.) 52:529–46.

24. Nill, K. R. 2006. *Glossary of Biotechnology and Nanobiotechnology Terms*. 4th ed. Boca Raton, FL: CRC Press.

25. Andrievski, R. A. 2005. Main problems in modern nanomaterials science. *Mater. Sci. Forum* 494:113–20.

26. Porschke, D. 1971. Cooperative nonenzymatic base recognition. II: Thermodynamics of the helix coil transition of oligoadenylic + oligouridylic acids. *Biopolymers* 10:1989–2013.

27. Bloomfield, V. A., D. M. Crothers, and I. Tinoco, Jr. 1974. *Physical Chemistry of Nucleic Acids*. New York: Harper & Row.

28. Grosberg, A. Yu. 1979. On some possible conformational states of homogeneous elastic polymer chain. *Biophysics* (Russian ed.) 24:32–37.
29. Grosberg, A. Yu., I. Ya. Erukhimovich, and E. I. Shakhnovich. 1981. On the theory of DNA compactization in polymeric solution. *Biophysics* (Russian ed.) 26:897–905.
30. Roco, M. C., S. Williams, and P. Alivisatos, ed. 2000. *Nanotechnology research directions: IWGN workshop report: Vision for nanotechnology R&D in the next decade.* Dordrecht, Germany: Springer.
31. International Agency for Research on Cancer. 1977. *IARC monographs on the evaluation of carcinogenic risk of chemicals to man.* Vol. 14, *Asbestos.* Lyon, France: IARC.
32. Magrez, A., S. Kasas, V. Salicio, et al. 2006. Cellular toxicity of carbon-based nanomaterials. *Nano Lett.* 6:1121–25.
33. Fenoglio, I., M. Tomatis, D. Lison, et al. 2006. Reactivity of carbon nanotubes: Free radical generation or scavenging activity? *Free Radical Biol. Med.* 40:1227–33.
34. Colvin, V. L. 2003. The potential environmental impact of engineered nanomaterials. *Nat. Biotechnol.* 21:1166–70.
35. Hoet, P. H. M. 2004. Health impact of nanomaterials? *Nat. Biotechnol.* 22:19.
36. Oberdorster, G., E. Oberdorster, and J. Oberdorster. 2005. Nanotoxicology: An emerging discipline evolving from studies of ultrafine particles. *Environ. Health Perspect.* 113:823–39.
37. Lam, C., J. T. James, R. McCluskey, S. Arepalli, and R. L. Hunter. 2006. A review of carbon nanotube toxicity and assessment of potential occupational and environmental health risks. *Crit. Rev. Toxicol.* 36:189–217.
38. Muller, J., F. Huauxa, N. Moreau, et al. 2005. Respiratory toxicity of multi-wall carbon nanotubes. *Toxicol. App. Pharmacol.* 207:221–31.
39. Singh R., D. Pantarotto, L. Lacerda, et al. 2006. Tissue biodistribution and blood clearance rates of intravenously administered carbon nanotube radiotracers. *Proc. Natl. Acad. Sci. USA* 103:3357–62.
40. Aldobaev V. N., L. A. Eremenko, A. A. Mazanova, et al. 2008. The study of the distribution, excretion and assessment of the main pharmacokinetic parameters of single-walled carbon nanotubes (SWCNT) in the body of small laboratory animals with different methods of administration (Russian ed.). Paper presented at the Nanotechnology International Forum, Moscow, December 3–5.
41. Manna, S. K., S. Sarkar, J. Barr, et al. 2005. Single-walled carbon nanotube induces oxidative stress and activates nuclear transcription factor-KB in human keratinocytes. *Nano Lett.* 5:1676–84.
42. Kroto, H. W., J. R. Heath, S. C. O'Brien, R. F. Curl, and R. E. Smalley. 1985. C60: Buckminsterfullerene. *Nature* 318:162–65.
43. Sayes, C. M., J. D. Forter, W. Guo, et al. 2004. The different cytotoxicity of water-soluble fullerenes. *Nano Lett.* 4:1881–87.
44. Jia, G., H. Wang, L. Yan, et al. 2005. Cytotoxicity of carbon nanomaterials: Single-wall nanotube, multi-wall nanotube, and fullerene. *Environ. Sci. Technol.* 39:1378–83.
45. Papageorgiou, I., C. Brown, R. Schins, et al. 2007. The effect of nano- and micron-sized particles of cobalt-chromium alloy on human fibroblasts in vitro. *Biomaterials* 28:2946–58.
46. Jin, Y., S. Kannan, M. Wu, and J. X. Zhao. 2007. Toxicity of luminescent silica nanoparticles to living cells. *Chem. Res. Toxicol.* 20:1126–33.
47. Lovrić, J., S. J. Cho, F. M. Winnik, and D. Maysinger. 2005. Unmodified cadmium telluride quantum dots induce reactive oxygen species formation leading to multiple organelle damage and cell death. *Chem. Biol.* 12:1227–34.
48. Dubertret, B., P. Skourides, D. J. Norris, V. Noireaux, A. H. Brivanlou, and A. Libchaber. 2002. In vivo imaging of quantum dots encapsulated in phospholipid micelles. *Science* 298:1759–62.

49. Hardman, R. 2006. A toxicologic review of quantum dots: Toxicity depends on physico-chemical and environmental factors. *Environmental Health Perspective* 114:165–72.
50. Wang, X. B., H. Y. Gao, B. L. Hou, J. Huang, R. G. Xi, and L. J. Wu. 2007. Nanoparticle realgar powders induce apoptosis in U937 cells through caspase MAPK and mitochondrial pathways. *Arch. Pharm. Res.* 30:653–58.
51. Qi, L. F., Z. R. Xu, Y. Li, X. Jiang, and X. Y. Han. 2005. In vitro effects of chitosan nanoparticles on proliferation of human gastric carcinoma cell line MGC803 cells. *World J. Gastroenterol.* 11:5136–41.
52. Reeves, J. F., S. J. Davies, N. J. Dodd, and A. N. Jha. 2008. Hydroxyl radicals (OH) are associated with titanium dioxide (TiO2) nanoparticle-induced cytotoxicity and oxidative DNA damage in fish cells. *Mutat. Res.* 640: 113–22.
53. Jin, Y., S. Kannan, M. Wu, and J. X. Zhao. 2007. Toxicity of luminescent silica nanoparticles to living cells. *Chem. Res. Toxicol.* 20:1126–33.
54. Borisov, N. M. 2008. Nanoparticles nanopreparations and blood-brain barrier. Paper presented at the International Forum on Nanotechnology, Moscow, December 3–5.
55. Wang, J. J., B. J. Sanderson, and H. Wang. 2007. Cyto- and genotoxicity of ultrafine TiO_2 particles in cultured human lymphoblastoid cells. *Mutat. Res.* 628:99–106.
56. Wang, L., J. Mao, G.-H. Zhang, and M.-J. Tu. 2007. Nano-cerium-element-doped titanium dioxide induces apoptosis of Bel 7402 human hepatoma cells in the presence of visible light. *World J. Gastroenterol.* 13:4011–14.
57. Gurr, J. R., A. S. Wang, C. H. Chen, and K. Y. Jan. 2005. Ultrafine titanium dioxide particles in the absence of photoactivation can induce oxidative damage to human bronchial epithelial cells. *Toxicology* 213:66–73.
58. Warheit, D. B., R. A. Hoke, C. Finlay, E. M. Donner, K. L. Reed, and C. M. Sayes. 2007. Development of a base set of toxicity tests using ultrafine TiO_2 particles as a component of nanoparticle risk management. *Toxicol. Lett.* 171:99–110.
59. Lippmann, M. 1994. Nature of exposure to chrysotile. *Ann. Occup. Hyg.* 38:459–67.
60. Hoet, P. H. M., I. Bruske-Hohfeld, and O. V. Salata. 2004. Nanoparticles: Known and unknown health risks. *J. Nanobiotechnol.* 2:1–15.
61. Salata, O. V. 2004. Application of nanoparticles in biology and medicine. *J. Nanobiotechnol.* 2:1–6.
62. Sycheva, L. P. 2008. Assessment of genetic safety of nanomaterials. Paper presented at International Forum on Nanotechnology, Moscow, December 3–5.
63. Organisation for Economic Co-operation and Development. 2005. *Report of the OECD workshop on the safety of manufactured nanomaterials.* No. 1, Series on the Safety of Manufactured Nanomaterials. Paris: Organisation for Economic Co-operation and Development.
64. Organisation for Economic Co-operation and Development. 2007. *Current developments/activities on the safety of manufactured nanomaterials.* No. 29, Series on the Safety of Manufactured Nanomaterials. Paris: Organisation for Economic Co-operation and Development.
65. Robichaud, C. O., D. Tanzil, U. Weilenmann, and M. R. Wiesner. 2005. Relative risk analysis of several manufactured nanomaterials: An insurance industry context. *Environ. Sci. Technol.* 39:8985–94.
66. Ministry of Economic Development. 2007. Resolution of the Chief State Sanitary Doctor of the Russian Federation. No. 79, *On approval of the concept of toxicological studies, risk assessment methodology, identification and quantification of nanomaterials* (Russian ed.). Moscow, October 31.
67. Ilyin, L. A., and V. Yu. Soloviev. 2007. Key issues in nanotoxicology. In *Methodological problems in the study and evaluation of bio- and nanotechnology (nanowaves, particles, structures, processes, biological objects) in human ecology*

and environmental hygiene (Russian ed.), ed. Yu. A. Rakhmanin. Moscow: Russian Academy of Medical Sciences, and the Ministry of Health and Social Development of the Russian Federation.

68. Onishchenko, G. G., A. I. Archakov, V. V. Bessonov, et al. 2007. Key issues in nanotoxicology. In *Methodological problems in the study and evaluation of bio- and nanotechnology (nanowaves, particles, structures, processes, biological objects) in human ecology and environmental hygiene* (Russian ed.), ed. Yu. A. Rakhmanin. Moscow: Russian Academy of Medical Sciences, and the Ministry of Health and Social Development of the Russian Federation.

2 Nanostructures Formed by Hybridization of Synthetic Single-Stranded DNA Molecules

2.1 GENERAL CONCEPT OF THE FUNDAMENTAL PROPERTIES OF DNA USED IN NANOTECHNOLOGY

In chapter 1, it was noted that natural evolution has resulted in the creation of such molecules as nucleic acids, proteins, and their assemblies, whose properties and functions are only now being simulated by scientists. For instance, ribosome particles with a size of 20 nm are supramolecular machines spontaneously assembled from more than fifty individual proteins and nucleic acids, which illustrates the power of biologically programmed molecular recognition.

Restructuring Figure 1.8 in chapter 1, it can be noted that there is an area (Figure 2.1) on the size scale where an intensive transfer of information from biological and chemical science to engineering nanotechnology can take place [1]. Figure 2.1 shows that, on the one hand, molecules and structures synthesized in natural or artificial conditions using the bottom-up technology have a typical size varying from 5 to 500 nm. On the other hand, the size of microprocessors constructed by the top-down technology currently reaches 200 nm. Though these devices can be created using various top-down techniques, such as photolithography, such techniques will scarcely make it possible to start a large-scale production of parts with a size below 100 nm in the near future. It is obvious that the creation of practically significant structures at the nanometer scale is one of the key challenges of science and technology in the twenty-first century [2]. This means that commercial demand for the production of miniature devices will determine the need to develop building blocks based on concept and principles that are used in nature to create nanosized biosystems. At the same time, the building blocks must have a size that fills the gap between the submicrometric sizes reached by the classical top-down technology and such sizes that can be achieved using the classical bottom-up technology, such as chemical synthesis or biological self-assembly.

Analysis of the fundamental backgrounds of physical chemistry, biochemistry, and synthetic chemistry of biological molecules, considering the requirements for nanostructured objects enumerated in chapter 1 and the data given in Figure 2.1,

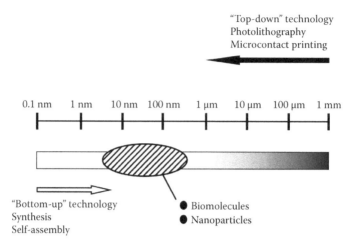

FIGURE 2.1 The area of intercrossing of different sciences in nanotechnology: the area, where the most active transfer of knowledge from biology, chemistry, etc. into nanotechnology as expected, is marked.

leads to the conclusion that, due to their geometric size, two groups of compounds are the most suitable for the creation of nanosized structures:

1 Biopolymers (such as nucleic acids and proteins)
2 Nanoparticles (nanoclusters) of various origin [1]

The formation of complex spatial objects with regulated properties (nanostructures, nanobiomaterials, nanoconstructions) using nucleic acid molecules and their complexes as construction blocks is called *nucleic acid based nanodesign* or structural nucleic acid nanotechnology [3–5]. At this time, there are two different strategies of nanodesign:

1. Creation of nanostructures using single-stranded polynucleotide (DNA) molecules containing deliberately selected sequences of nitrogen bases
2. Creation of nanoconstructions using spatially fixed linear double-stranded DNA molecules (or their complexes)

This chapter examines various approaches to the formation of nanostructures based on single-stranded nucleic acid molecules. This area of research started to form in the beginning of the 1990s as a result of pioneering work by Seeman [4], and significant progress had been achieved by 2003. Within this period, significant results have been reported, and the immutable principles of the formation of DNA nanostructures have been established [5].

Before describing the properties of DNA nanostructures, it is necessary to examine the fundamental properties of these molecules that determine the possibility of their application in nanotechnology [6].

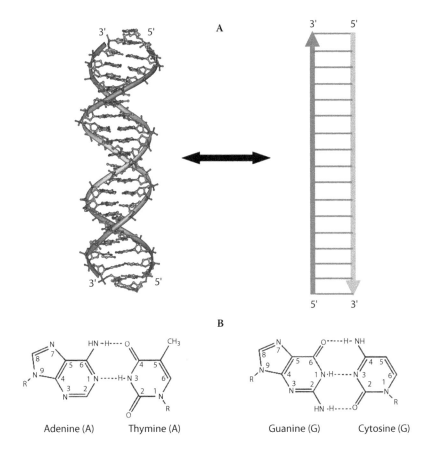

FIGURE 2.2 Scheme of a double-stranded DNA molecule consisting of two hydrogen bonded opposite polynucleotide chains (marked red and green). (A) A simplified structure is depicted as a ladder; the arrows show the 5′-3′-direction (polarity) of the polynucleotide chains; (B) complementary base pairs (A and T), (G and C) that form specific Watson–Crick hydrogen bonds.

It is known that chromosomes of all living systems include DNA molecules; the genetic code is encrypted in their structures. In 1953, James Watson and Francis Crick proposed a double-helical model of the DNA molecule (Figure 2.2(A)). A DNA molecule (Figure 2.2(A)) is a linear polymer consisting of two sugar-phosphate chains that contain nucleotides—nitrogen bases (Figure 2.2(B)): adenine (A), guanine (G), cytosine (C), and thymine (T) bonded to sugar residues [7]. Two opposite sugar-phosphate chains twist upon each other and form a linear double spiral (double helix). Each chain contains complementary base pairs, such as A-T (with two hydrogen bonds between them) and G-C (with three hydrogen bonds between them) (Figure 2.2(B)). The complementarity concept entails the formation of hydrogen bonds between A and T, as well as between G and C, and an accurate steric conformity between the surfaces of nitrogen base pairs, and, consequently, the oligonucleotide chains that include these nitrogen bases (including their mutual conformation adjustment) [8].

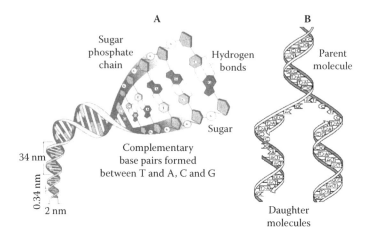

FIGURE 2.3 Model of double-stranded DNA molecule: (A) a size of a helical turn (3.4 nm) and the diameter of the helical structure (2 nm) are marked; (B) the scheme illustrates the formation of "daughter" molecules as a result of the untwisting of the "parent" DNA molecule.

Negatively charged groups of phosphoric acid residues provide the electrostatic repulsion between chains in a DNA molecule and give the molecule the properties of an acid soluble in standard water–salt solutions. Figure 2.2 also contains a schematic image of a DNA molecule as two straight lines with opposite polarity. Such images will be used to further illustrate the material. The simplicity and elegance of the Watson–Crick model has not only revolutionized genetics, but also exerted a great influence on other areas of science, from law to medicine. This discovery has answered the question of how genetic information is stored and how it is interpreted and processed in a DNA molecule. It should be emphasized once again (Figure 2.3(A)) that a DNA double helix is a nanometric object [9]. The diameter of a molecule is about 2 nm, and the distance between pairs is 0.34 nm. Every helical turn consists from 10 to 10.5 base pairs per turn, so the total size of one turn is about 3.5 nm.

The spatial structure of a double-stranded DNA molecule depends on the base sequence: The differences in spatial organization caused by the presence of different pairs in the chain lead to changes in the relative orientation of mean planes of the base pairs. The mutual orientation of base pairs and their orientation to the axis of a linear DNA molecule [10] are expressed in such terms as *base slope angle*, *base pair helical twist angle*, and *inclination angle of base pairs* (Figure 2.4). These angles determine both the local bending of a DNA chain and the total curvature of a DNA molecule. The shape taken by an individual DNA molecule in space and time can be analyzed in terms of superposition of thermal fluctuations of the structure and the inner low-energy structure typical of the sequence of nitrogen bases [10]. In solution, a DNA molecule exists in a canonical B form, as a rule, although, depending on the base sequence and properties of the solvent, there may be a wide range of DNA structural forms.

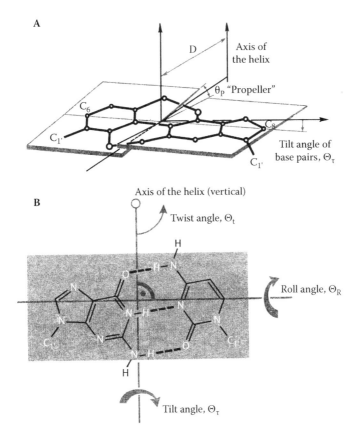

FIGURE 2.4 The parameters that determine the orientation of individual nitrogen bases (A) and base pairs (B) in respect to the axis of a linear DNA molecule.

The study of the structural peculiarities of DNA can be started by considering, first of all, the specific properties of nitrogen base sequences in a DNA chain. The first specific sequence is the so-called palindromic sequence or palindrome repeat (Figure 2.5) [11]. (A palindrome is a word, phrase, or sentence that reads the same from the left to the right and from the right to the left, such as "rotator" or "nurses run.") For a DNA molecule, this term means that both DNA chains contain sequences connected by a twofold symmetry (Figure 2.5). In order to superimpose one repeat on the other, it must be rotated 180° around the horizontal axis and then again about the vertical axis (shown by the colored arrows).

A mirror repeat of nitrogen bases has a symmetric sequence on each strand. Superimposing one repeat on the other requires only a single 180° rotation about the vertical axis (Figure 2.5).

Palindromic sequences of DNA or RNA can come from alternative structures where bases form pairs in one chain. When only one DNA chain is included in the process, a hairpin structure is formed (Figure 2.5(A)). When both of the DNA chains take part in the process, the structure is called a cruciform (Figure 2.5(B)).

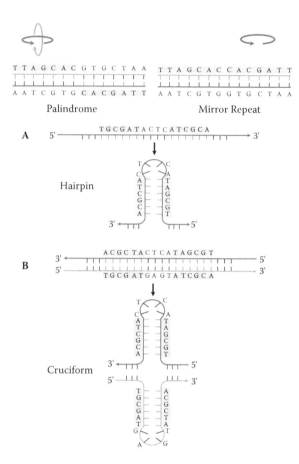

FIGURE 2.5 Nitrogen base sequences in DNA strand (repeats) that determine the possible existence of various local structural forms. The base sequences that can form complementary pairs with bases of the same or the opposite strand are marked in color. (A) Formation of a hairpin; (B) formation of a cruciform DNA structure.

Palindromic sequences are observed in DNA molecules with high molecular mass and may contain thousands of base pairs. It is not known how many molecules in various cells include a cruciform structure, though the existence of such structures has been observed in *E. coli* cells.

A double-stranded DNA structure is given again in Figure 2.6(A) (the strands are numbered 1 and 2). The key feature of a DNA molecule that is often ignored is the fact that the axis of a double helix is linear, i.e., it is not branched [5, 12]. This is significant because the structure of a fragment complementary to one of the DNA chains can be calculated. Indeed, chain 1 in the depicted structure is complementary to 2, as it contains nucleotides that form Watson–Crick hydrogen bonds with nucleotides in chain 2. If one sugar-phosphate bond in one of the chains of the initial DNA molecule is severed, which is called a "nick," such molecule is called a "nicked molecule" and it looks as shown in Figure 2.6(B). In this case, the initial

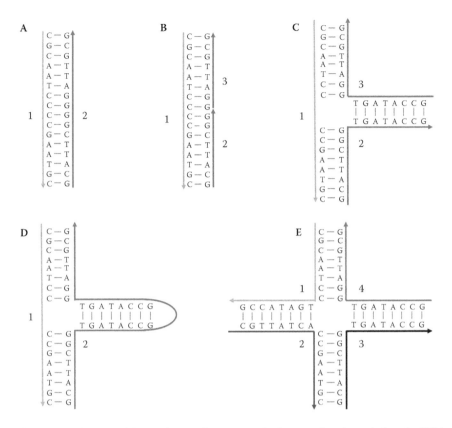

FIGURE 2.6 Structural forms that results as one or both sugar phosphate chains of a DNA molecule are broken. (Courtesy of Nadrian C. Seeman.)

chain 2 (Figure 2.6(A)) is broken into two single-stranded fragments (2 and 3), and the sum of 2 and 3 is complementary to chain 1. With the help of an enzyme, the break between chains 2 and 3 can be eliminated to obtain the initial DNA structure (Figure 2.6(A)). Moreover, the nucleotides can be added to the 3′-end of chain 2 and the 5′-end of chain 3.

Figure 2.6(C) shows what happens if lengthened nucleotide sequences are complementary. In this case, the initial chain 2 is branched, while the horizontal branching chain may have any length. The received structure is a three-arm DNA branched junction. It contains a complex of chains 2 and 3 that is also complementary to chain 1. (The replication of this DNA structure with the help of polymerase may not be very efficient, though it may lead to the formation of complementary chains.) Chains 2 and 3 can also be connected with a single-chain hairpin containing four or five thymine nucleotides (Figure 2.6(D)). If an extra break is made in chain 1 in Figure 2.6(C), and then the formed ends are lengthened like in the transition from Figure 2.6(B) to Figure 2.6(C), another junction point can be introduced into the structure. As a result, the four-arm DNA branched junction shown in Figure 2.6(E) is created. At the transition from Figure 2.6(D) to Figure 2.6(E) (the numbers of complementary chains have been changed), it can be seen that the set of the three other chains is complementary to any chain. For example, a complex of chains 2, 3, and 4 is complementary to chain 1, though there is no direct interaction between chains 1 and 3. It seems that the uniqueness of the complementarity is broken. In Figure 2.6(A), chain 2 was a unique fragment complementary to chain 1. In Figure 2.6(E), chain 3 becomes complementary to chain 1, though it is different from chain 2. Actually, the general principles of complementarity are not violated, and the example just demonstrates that any nucleotide fragments can be introduced between any two base pairs of a chain if the interaction takes place between Watson–Crick base pairs [12].

Figure 2.6 allows one to make the following statement: DNA nitrogen bases have peculiar properties. First, their spatial structure provides the possibility of forming complementary pairs; second, the effective recognition is only typical of complementary base pairs; and, finally, a stable structure that formed by complementary base pairs does not depend much on exterior conditions. An important consequence of the enumeration of base properties is that the formation of complementary Watson–Crick base pairs requires an intact base pair structure. Any displacement of electron or steric structure of base pairs may reduce their ability to form Watson–Crick pairs, and, consequently, different spatial forms of DNA molecules with significantly different properties correspond to different base structures.

The complementarity determines the modern approach to consideration of the role of nucleic acid molecules in molecular genetics [13]. Indeed, according to the complementarity principle, daughter molecules with the same nucleic acid complementarity as the parental molecule (Figure 2.3(B)) can be received by the effect of a semiconservative mechanism. The obtained DNA molecules will be stable under certain conditions, such as temperature and fixed properties of the solution. To create DNA daughter molecules, complementary nitrogen bases are synthesized on single-stranded DNA fragments and then joined into two continuous daughter polynucleotide chains (Figure 2.3(B)).

Double-stranded DNA molecules may have single-stranded fragments at their ends (extensions or "sticky ends") that can form complementary pairs with alien single-stranded DNA fragments (Figure 2.7(A)). The most important feature of the sticky ends is that they provide the specificity of the interaction between end nucleotide sequences of single-stranded DNAs [14]. The useful peculiarity of such interaction is that when the sticky ends overlap, specific hydrogen bonds are formed between single-stranded DNA base pairs. Figure 2.7(B) shows that under proper conditions, the sticky ends of two DNA molecules recognize and bind to each other specifically by hydrogen bonding. (There are two key features to sticky-ended cohesion that make it important for the application in DNA nanotechnology, i.e., predictable affinity and structure.) The further treatment of the received structure that contains two sugar-phosphate chain breaks by proper enzymes (ligases) leads to the formation of a rigid, helical, double-stranded DNA molecule (B-form DNA) with complementary base pairs (Figure 2.7(C)). As base sequences in the molecules in Figure 2.7(A) may be known, the example proves that in the case of DNA, intermolecular interactions can be predicted and programmed.

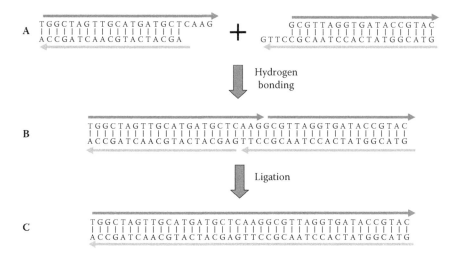

FIGURE 2.7 "Sticky ended" cohesion and ligation. (Courtesy of Nadrian C. Seeman.)

A more complicated process, namely, the formation of a double-stranded DNA molecule from two single-stranded molecules with complementary base pairs, is known (Figure 2.8(A)). Here, each interacting unit (oligomer) is an array of four primary nitrogen bases that selectively bind other oligomers in only two possible pairs. As a result, a strong and specific double-stranded complex is formed between two oligomers. But several incorrect complexes could also be formed (Figure 2.8(B)). Single mismatches can be relatively stable. The process of creating a double-stranded molecule from two joining single-stranded complementary DNA chains is called *hybridization* [15].

In the case of DNA, the hybridization is a thermodynamically regulated and reversible process. While a double-stranded DNA molecule is separated into two single-stranded chains at the increase in temperature, two complementary single-stranded

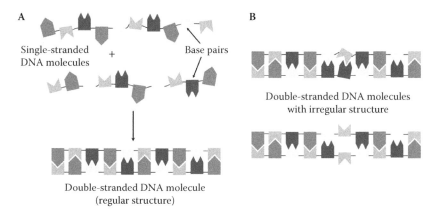

FIGURE 2.8 **(See color insert)** Formation of a double-stranded DNA molecule as a result of the complementary recognition mechanism of nucleation-zipping.

molecules are joined at the decrease in temperature. Another circumstance that has an important meaning for DNA nanotechnology can be highlighted: A single-stranded DNA chain structure corresponds to the structure of flexible-chain polymers with most accidental orientation of neighboring nitrogen bases, while the double-stranded DNA molecule formed as a result of hybridization is a rigid molecule with fixed location of nitrogen bases in respect to the DNA molecule's axis.

Two parameters play an important role in the process of hybridization: the affinity between nitrogen base pairs and the specificity of the interaction between them. Analysis shows that there is a wide range of conditions at which the formation of incorrect sequences (Figure 2.8(B)) is possible. For an effective self-assembly, the hybridization of single-stranded DNA molecules must be as high as possible. This means that the formation of base pairs must take place at a maximum efficiency, while the percentage of wrong base pairs has to be minimal. Unfortunately, a simple optimization of the affinity, for example, by increasing the length of the molecule and the content of G- and C-nitrogen bases, does not necessarily lead to high specificity because the two parameters do not correlate at the hybridization of nucleic acids [8]. The stability of undesirable hybrids measured, for example, by the change in the free energy at the formation of complementary pairs has to be minimized, while the stability of the desirable pairs must be maximal. This means that nitrogen bases in the initial DNA chains must be accurately chosen (synthesized), and the experimental conditions for the hybridization must be selected to provide both the affinity and the specificity of the hybridization. The optimal option is to use deliberately designed oligonucleotides that contain sequences of nitrogen bases with a specified structure.

From this discussion, it can be concluded that the possibility of using DNA molecules as building blocks—in the case of the hybridization technique of DNA nanostructure formation—is determined by a number of unique properties of these molecules.

1. Nitrogen bases can form specific (complementary) pairs; the complementarity of bases is the first bedrock of DNA structural nanotechnology.
2. Flexible single-stranded fragments of nucleic acid molecules with a specified base sequence (the sticky ends) hybridize with complementary fragments and from rigid double-stranded molecules whose local structure corresponds to the B form. The hybridization is the second bedrock of DNA nanotechnology.
3. The predictability of the location of all atoms in the structure of a helical double-stranded molecule relative to its linear axis, as well as the properties of rigid linear DNA molecules and the nature of intermolecular interactions under different conditions, is the third bedrock of DNA nanotechnology.
4. The synthesis of DNA molecules of different sizes and chemical composition (length up to several microns) in reasonable quantities with the help of modern chemical (biochemical) methods is the fourth bedrock of DNA nanotechnology.

Hence, by choosing the right sequences, one can "program" the interactions between DNA molecules and use them to generate a molecular network consisting of rigid and flexible fragments.

2.2 HYBRIDIZATION TECHNIQUE OF CREATING NANOSTRUCTURES BASED ON SYNTHETIC SINGLE-STRANDED DNA MOLECULES (BOTTOM-UP APPROACH TO THE CREATION OF DNA NANOSTRUCTURES)

The fundamental problem of using linear DNA molecules as building blocks in the bottom-up approach used to create DNA nanostructures is the following: Molecules with flat helical axes are logically equivalent to straight lines, so linear DNA molecules cannot form extended two- or three-dimensional (2-D or 3-D) structures with properties that attract the interest of researchers under standard solvent properties [14]. This means that the formation of an extended spatial nanostructure with hydrogen bonds between neighboring elements requires the introduction of junction points into the initial DNA structure that would work as the angles in the created construction [14].

From this point of view, the branched DNA structures in Figure 2.9 attract some interest. As shown in Figure 2.9(A), four complementary oligonucleotide molecules (1–4) can form a branched DNA structure. (Here: the place where the strands cross is the junction; the "migration of a junction point" corresponds to the movement of this point up or down.) The peculiarity of such a structure is that, due to the symmetry of base sequences, it is unstable, and due to migration of a junction point, it can easily change the spatial shape (Figure 2.9(B)). Moreover, branched nucleic acid structures with high symmetry of nitrogen base sequences quickly turn into linear

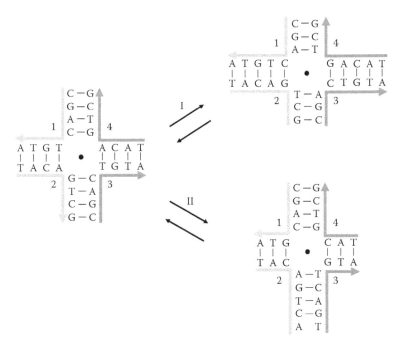

FIGURE 2.9 Migration of the branch point in a four-armed construction in either direction results in formation of two structures shown on the right. (Courtesy of Nadrian C. Seeman.)

double-stranded molecules as a result of migration of the junction point. However, the combination of sticky ends and branched motif of DNA molecules provides the possibility of the formation of a flat lattice.

Figure 2.10(A) gives a hypothetical example of the connection of a four-arm branched DNA molecule (cruciform structure) with sticky ends. (Here, the four sticky ends are labeled as X, Y and their complementary fragments as X′ and Y′.) The hybridization of four such molecules results—due to parallelism in the assembly process of four-arm DNA molecules with sticky ends [16]—in formation, initially, of a quadrilateral structure (Figure 2.10(B)). The formed structure also has sticky ends (open valences) on the outside. This means that at the necessary concentration of the four-arm branched DNA molecules in the solution (on the left), this motif could be assembled further into a 2-D periodic array (on the right). Under the action of DNA ligase, the breaks in sugar-phosphate chains at the sites of sticky ends joining in this structure can be "linked" (eliminated), and rigid double-stranded DNA (B form) ribs, connecting flexible branching points in periodic array, can be formed. The created structure (on the right) has a nanometric size. As the ribs in the square may be from 5 to 20 nm long, this kind of structure can be called a DNA nanostructure. As a result of the self-assembly process, beside the square, an infinite 2-D lattice can be formed. This assumption is based on the presence of sticky ends on the external edges of the structure, which makes the further assembly of a 2-D structure possible.

Nevertheless, the success of the assembly depends on both the stiffness of the DNA segments that form the edges of the square and the stability of junction points on the corners [14]. If any of these components is flexible, the square will not be the exceptional product, and the formation of a regular structure is highly unlikely (Figure 2.10). Though DNA molecules are considered to be flexible, they have a persistent length of about 50 nm at normal solvent properties, which means that locally a double-stranded DNA molecule is much more rigid. So, short DNA molecules with a length equal to two to three helical turns (6–10 nm) can be considered as rigid building blocks. Branched DNA molecules, shown in Figure 2.9, generate multiple cyclic products under the action of ligase, which demonstrates that the initial structures are movable, not rigid. At the same time, if a junction point in the construction shown in

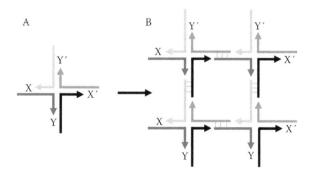

FIGURE 2.10 Formation of a two-dimensional lattice as a result of assembling of four-arm branched motif and sticky ends. (Courtesy of Nadrian C. Seeman.)

Figure 2.10 is also movable, the probability of the formation of a square structure is sharply reduced, and it becomes very important to select the sequences in the initial DNA molecules so that the position of the junction point in the branched structure can be fixed.

The key issue of the hybridization technique of DNA nanostructure creation is the search for and the synthesis of nitrogen base sequences that provide the formation of the desirable product and make it possible to avoid the formation of intermediate products of the assembly that interact with the final product. Basically, one can analyze the thermodynamic parameters of all of the possible sequences and choose a sequence and hybridization conditions that lead only to the desirable product. It has been empirically proven [4] that the approach called the "minimization of sequence symmetry" makes it possible to choose base sequences that are interesting for researchers. The first computer program for automatic search and formation of the nitrogen base sequences needed for the creation of nanoobjects—considering many of the requirements enumerated here—was proposed by Seeman [17] and called SEQUIN. Reviews of other programs can be found in the work of Brenneman and Condon [18].

Figure 2.11 shows a four-arm branched structure obtained by an approach called "sequence symmetry minimization." This figure illustrates that each arm of this structure is eight nucleotides long [5, 19, 20]. Each of the double helices contains 16 nucleotides divided into a series of overlapping elements (13 tetramers in this case). The branched structure composed of four DNA strands is labeled with Arabic numerals. An arrow marks each of the 3′-ends of the strand. Each strand is paired with two other strands to form double helical arms; the arms are numbered with Roman numerals.

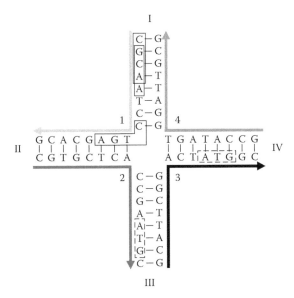

FIGURE 2.11 Sequence symmetry minimization produces a stable DNA branched structure. (Courtesy of Nadrian C. Seeman.)

Hydrogen bonds are shown as lines between corresponding base pairs. Base sequences in the structure are optimized to minimize the symmetry and non-Watson–Crick base pairing. Given the absence of symmetry in base pairs flanking the central branch point, this junction cannot undergo the branch migration isomerization reaction. Desirable sequences of nitrogen bases have been calculated by the computer program SEQUIN, and the synthesis of the initial single-stranded DNA molecules was based on the calculation. The formation of a four-arm DNA structure was conducted with the help of hybridization of the synthesized single-stranded DNA molecules using the following method.

Single-stranded DNA molecules were placed into a water–salt solution (40 mM TrisHCl; 20 mM acetic acid; 12.5 mM magnesium acetate). The solution was heated to 90°C and held at this temperature for 15 minutes, then cooled to 65°C (kept for 15 min.), 37°C (kept for 15 min.), to room temperature (22°C, kept for 20 min.), and to 4°C (kept for 20 min.). The phased cooling considers the difference in melting temperatures of A-T and G-C in base sequences, so it contributed to the increase in efficiency of hybridization of the corresponding sequences in single-stranded DNA molecules. After the received mixture was processed with T4 polynucleotide ligase and endonucleases HeaIII and PvuII, the properties of the final products were analyzed by electrophoresis in poly(acrylamide) gel. (Note that this technique and other similar methods were used in all further work to create nanometric objects from single-stranded DNA molecules.)

The method proposed by Seeman requires that every element, such as the boxed CGCA- and GCAA-sequences marked in Figure 2.11, be unique. Moreover, it is necessary that any element that spans a bend (such as the boxed CTGA) does not have its simple Watson–Crick complement (TCAG) anywhere in the sequence. With these constrains, the competition with the target octamers can come only from trimers, such as ATG, in the dotted boxes in Figure 2.11. This approach assumes that double helices are the most favorable structures that DNA molecules can form, and that maximizing double-helix formation will lead to the successful formation of branch points.

Therefore, the minimization of the sequence symmetry in individual DNA chains [21] makes it possible to design such single-stranded DNA molecules that will be hybridized in a solution and form a stable branched structure, i.e., a structure consisting of two, three, or more nucleic acid double helices located around a junction point (Figure 2.11). The success of this approach is based on the concept of the cooperativeness of the DNA double-helix formation as a result of hybridization of deliberately synthesized single-stranded molecules with specified base sequences.

The assumption that stable branched structures can be formed from short DNA molecules has led to another assumption, according to which branched nucleic acids can be considered as valence clusters and applied for structural engineering on a nanometric level. Using the high specificity of sticky ends joining catalyzed by ligase, branched structures can be joined in a specified way, forming closed structures. In such structures both edges and angles are nucleic acid double helices. Basically, various spatial shapes, from polyhedrons to periodic lattices like the one shown in Figure 2.10, and less regular structures may be formed.

The approach described here was developed to increase the efficiency of DNA molecule hybridization leading to the formation of DNA nanostructures. While this approach was being developed, there was an ongoing search for other DNA structural motifs with higher stiffness.

As noted in chapter 1, the success of nanotechnology depends on the efficient transfer of knowledge on the properties of biological structures formed as the result of evolution to actual experimental practice. During the search for more rigid DNA structures, the interest of researchers was attracted to branched DNA molecules formed as intermediate metabolites in biological processes, namely, the structure known as the Holliday junction [22].

The Holliday junction is the most prominent DNA intermediate in genetic recombination. It is known that genetic recombination takes place in living organisms from viruses to humans. Although DNA metabolism is connected to the stability of its structure, which is necessary to perform the role of a genetic material, the recombination is a process that requires mobility of the structure that provides the adaptation of individual DNA fragments to changing conditions [23]. The principal molecular peculiarity of the recombination is the interaction between two DNA fragments necessary for the formation of a new genetic material that may include segments of both of the molecules. The forming DNA may have such structural defects as deletions, insertions, sequence changes, restructurings, and specific marker exchange. The Holliday structure is the central element in the molecular mechanism of the process.

Figure 2.12 illustrates how the Holliday junction can take part in the genetic recombination. At the first stage (A), two homologous DNA molecules with different markers D-F (D'-F') and d-f (d'-f') line up. Then cuts in one strand of both DNA molecules takes place (B). The cut strands cross and join homologous strands to form a *Holliday structure* (or Holliday junction, C). It consists of four strands of DNA (C) that are paired into four double helical arms flanking a junction point. The Holliday junction contains two strands nicked and linked together again so that nucleotide sequences of the adjacent strands are joined. These strands form the junction (branch) point. The two other DNA strands are not changed and take no direct participation in the recombination process. The strands forming the junction point are called *crossover strands*, and the two other strands are called the *helical strands* [23]. The branch point typically is flanked by regions of dyad (homologous) nitrogen base sequence symmetry. This symmetry enables the branch point to relocate through an isomerization known as a branch point migration, which includes the change of the initial location of the junction point (D). The area where the strands of DNA molecules are joined is called a *heteroduplex region.*

Besides the migration of the junction point, another process may take place, namely, the crossover isomerization reaction. In this reaction, the crossover strands and the helical strands change places. Therefore, the crossover strands become the helical strands, while the helical strands that were initially intact take the place of the crossover strands. Attention should be paid to the fact that the Holliday junction is actually a four-arm branched DNA molecule. Cutting of the Holliday structure in vertical or horizontal (V and H) direction by enzymes and the further ligase processing of the obtained fragments lead to the formation of products with different properties (Figure 2.12(F) and 2.12(G), respectively). In the structure F, some markers change places. In the structure G, markers do not change their positions.

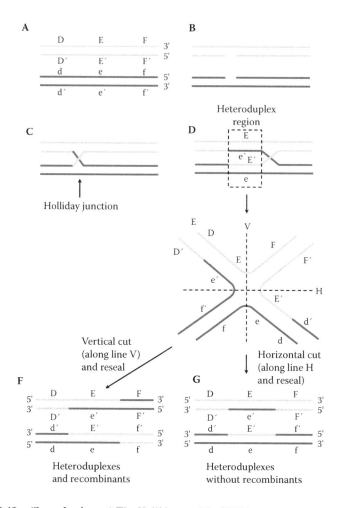

FIGURE 2.12 (**See color insert**) The Holliday model of DNA crossover.

The detailed mechanism of the genetic recombination was determined mostly by research using *E. coli* cells. In these cells, the recombination is initiated by RecBCD enzyme consisting of three subunits: RecB, RecC and RecD. The enzyme has both helicase and nuclease activity. First it uses its helicase activity to split the strands of neighboring DNA molecules (stage B in Figure 2.12). When it reaches the GCTGGTGG sequence called the Chi-site, the split strand is decomposed by the nuclease activity. The reason why the site is called the Chi-site is that the Greek letter χ looks like the junction point. The Chi-site corresponds to the point where the formation of the Holliday structure begins (stage C, Figure 2.12). After the DNA strands are split, RecBCD and RecA proteins are twisted around single-stranded DNA chains and initiate their transition to form the Holliday junction (stage D, Figure 2.12). The migration of the junction point is catalyzed by RuvA and RuvB enzymes. The final Holliday structure is also split by enzymes, namely, RuvC.

The most important Holliday junction peculiarity especially interesting for nano-technology is the crossover in DNA strands that is supposed to lead to the increase in the stiffness of the structure. However, it follows from Figure 2.12 that the natural Holliday junction is formed between two molecules with similar base sequences. Such sequence symmetry destabilizes the location of the junction point so that it can relocate freely. It is most difficult to study Holliday structures with physical methods because of the migration of the junction point, and, consequently, the high mobility of the structure.

However, the interest was attracted to the synthetic analogues of the natural Holliday structure, the four-arm branched structures. Examples of such structures can be found in Figure 2.13 [23]. The structure in the center of the figure is formed as the result of hybridization of four single-stranded DNA chains consisting of sixteen nucleotides:

The first strand (direction from 5′ to 3′) → C-G-C-A-A-T-C-C-T-G-A-G-C-A-C-G

The second strand (direction from 5′ to 3′) → C-G-T-G-C-T-C-A-C-C-G-A-A-T-G-C

The third strand (direction from 5′ to 3′) → G-C-A-T-T-C-G-G-A-C-T-A-T-G-G-C

The fourth strand (direction from 5′ to 3′) → G-C-C-A-T-A-G-T-G-G-A-T-T-G-C-G

The structure (center) is shown in a fourfold symmetric arrangement. The strand numbering is indicated by Arabic numerals, and the arms are numbered by Roman numerals. Each strand takes part in the formation of a double-helical arm; each of the arms consists of eight complementary nucleotide pairs. The physical parameters of the obtained molecule have been determined by various methods. The decomposition of the molecule under the action of free radicals has shown a structural twofold symmetry in the solution corresponding to the conformation in which the helical domains are joined in pairs. As seen in Figure 2.13, the four-arm structure can fold and form two helical domains (stacking arm I on arm II, and arm III to arm IV, results in two helical domains). Both of the possible coplanar arrangements of these helical domains with the antiparallel (on the left) and the parallel arrangement (on the right) are shown. In Figure 2.13(B) are shown two conformationally constrained junctions (on the left, the stacking domains are antiparallel, on the right they are parallel), which contain identical base-paired arms. The loops consist of a $d(T)_9$ sequence.

The structure in Figure 2.13 is basically a synthetic analogue of the natural Holliday junction. The application of various synthetic methods makes it possible to create branched molecules with fixed junction point. Namely, the two structures in Figure 2.13 contain thymine loops between the strands that fix the junction points [24].

Despite the success in the creation of branched structures with the help of the method described here, a technique based on the theoretical calculation and syn-thesis of single-stranded molecules with no symmetry in the nucleotide sequences has become more popular (see Figure 2.11). This technique results in the formation of branched molecules with a fixed junction point [5]. A synthetic four-arm DNA molecule where each arm has a unique sequence is a stable analogue of the natural

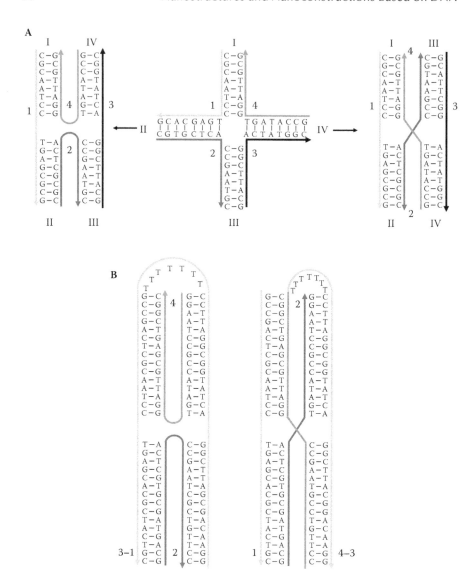

FIGURE 2.13 The schematic representation of various oligonucleotide junctions. (Courtesy of Nadrian C. Seeman.)

Holliday junction, where the migration of the junction point is suppressed. Various kinds of such molecules have been created and thoroughly studied [23–25].

It is obvious that the formation of sticky ends in DNA molecules with a fixed junction point may lead to the formation of structures that can hybridize, which makes products of the ligase processing more complicated in comparison to the products received from linear DNA. However, periodic structures can be obtained from such products. Because the diameter of a DNA molecule is about 2 nm, the assembly of nanometric DNA structures can be formed.

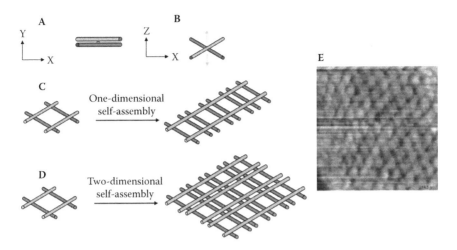

FIGURE 2.14 (**See color insert**) Lattices built from branched junctions. (Courtesy of Nadrian C. Seeman.)

One of the possible variants of formation of a periodic structure [16] using the Holliday junction motif is shown in Figure 2.14. The Holliday junction is shown in Figure 2.14(A), where the dyad axis is labeled by a small dot. Figure 2.14(B) shows a view with the dyad axis vertical. The blue helical domain is rotated 30° about the vertical, so that its right end penetrates the page [25]. The helical domains are antiparallel and form a right-hand twisted structure with a twist angle around 60° [25]. This provides the possibility to form a two-layer structure where each of the layers is formed from the corresponding domains. Although such a structure remains flexible, a more rigid construction can be created by joining the four Holliday junctions (Figure 2.14(C,D)) into a rhombus or a parallelogram [26]. This structure (Figure 2.14(C)) consists of two parallel edges in one plane and another pair of edges located in the other plane at a distance of approximately 2 nm from the first pair. The junction points in the structure must be coplanar. The helical domains in Figure 2.14(C) contain six helical turns each; there are four turns between the junction points; and there is one turn between the end of each molecule and a junction point. The creation of sticky ends (Figure 2.14(C)) in the helical domains makes it possible for the structures to hybridize and form one-dimensional or two-dimensional lattices. One-dimensional self-assembling of this DNA motif is accompanied by the formation of a railroad track–like array (Figure 2.14(C)) with helices located at different distances of two or four helical twists (the product is called a 4+2 motif). The size of a cell can also be changed to create lattices with 6×4, 6×2, 4×2, and 2×2 motifs. If all of the helices have complementary sticky ends, a 2-D array built from DNA parallelograms shown in Figure 2.14(D) can be formed.

Using the computer program SEQUIN, single-stranded DNA molecules with different lengths (from 6.3 to 12.1 nm) and specified nitrogen base sequences providing the targeted hybridization have been designed. As a result of hybridization under the conditions close to those described previously, with further ligase processing of the received products, flat nanostructures have been formed from these molecules. Figure 2.14(E) gives the image of such structure taken by atomic force microscopy (AFM) [27].

Therefore, structures with increased stiffness can be formed from synthetic DNA molecules and used as building blocks for hybridization to form nanostructures. However, a detailed analysis has shown that the stiffness of structures like the Holliday structure is insufficient to form nanostructures with high practical yield.

That is why attention was paid to another biological hint, i.e., to structure existing in biology, namely, a DNA molecule with double crossover known as an intermediate structure that is formed during meiosis. The DNA double-crossover molecule contains two crossover links between helical domains. The double-crossed (DX) molecules consist of two double-stranded helices that exchange individual chains in two crossover points. These kinds of molecules are different in the mutual orientation of the helical domains, the continuity of the sugar-phosphate chains, as well as the number of twists of the double helix between the junction points [7]. Such a DNA motif [28] and the nucleotide composition of some theoretically selected and then synthesized single-stranded molecules that can form double-crossover structures as the result of hybridization [7, 29, 30] are shown schematically in Figure 2.15.

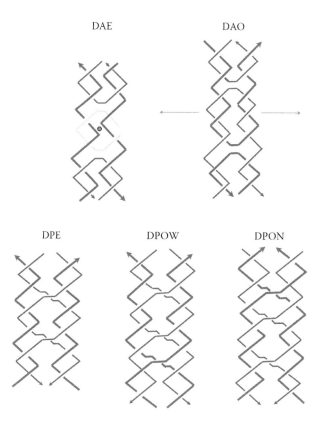

FIGURE 2.15 **(See color insert)** Five various types of DNA double-crossover molecules. (Courtesy of Nadrian C. Seeman.)

There are five different DX patterns. In Figure 2.15, DNA strands are shown as zigzags, where two consecutive perpendicular lines correspond to a full helical turn of a strand. The strands are marked with different colors for better orientation. The 3′-ends of the strands are marked with arrows. The structures shown in Figure 2.15 are called by the first letters of their full names. All the names have the letter *D* marking the double-crossover [31]. The second letter corresponds to the orientation of the two double-stranded domains: A = antiparallel; P = parallel. The third letter shows the number of helical half-turns between junction points: E = an even number; O = an odd number. The fourth letter is used to describe the parallel double-crossover molecules with an odd number of helical half-turns between the junction points. An extra half-turn may correspond to the distance between the major grooves (wide separation, W) or the minor grooves (narrow separation, N). The isomers of an antiparallel DX molecule known as DAE and DAO are shown in the top part of the Figure 2.15. The creation of these structures depends on the helical twists of B-form DNA, where a full turn takes place in every 10.5 base pairs.

DAE is one of the isomers of an antiparallel DX molecule with an even number of half-turns of the double helix between the junction points, for instance, four half-turns, i.e., twenty-one base pairs. A DAE molecule contains five strands: Two of them are continuous helical strands marked with red, and three of them are the crossover strands marked with blue and green, including the green strand in the center.

DAO is another isomer of an antiparallel DX molecule with an odd number of helical half-turns between the junction points, for instance, three half-turns, i.e. sixteen base pairs. A DAO molecule only contains four strands. Two of the strands are marked with red, the other two are blue.

Therefore, DAE and DAO patterns have different numbers of helical half-turns between the junction points. The symmetry axis is perpendicular to the page plane in the case of DAE (the symmetry is changed by the nick in the central strand) and located horizontally in the page plane in the case of DAO. For DAO, strands marked with different colors are symmetric.

The bottom part of Figure 2.15 shows three variants of parallel double-crossed molecules. The three parallel double-crossed molecules are different by the number of helical half-turns between junction points: an even number for DPE or an odd number for DPOW and DPON. DPOW and DPON are different by the presence of an additional major or minor groove between the junction points. Formation of a DPOW takes sixteen nitrogen base residues between junction points, while the formation of a DPON requires fourteen residues between the junction points.

If the helical domains are parallel, both major and minor grooves of both of the helices cross each other in every turn. Consequently, a strong Coulomb interaction (repulsion) takes place between two parallel strands, so these patterns have an unstable structure [29, 32]. Two antiparallel patterns, where a minor groove of one helix corresponds to a major groove of the other helix, are more stable because of the reduction of the Coulomb repulsion.

As a DX molecule can be considered as two four-arm structures joined with two adjacent arms, the migration of a junction point in these structures is possible. However, as two junction points have to relocate in concord, this process is less probable in comparison to the process that takes place in regular branched DNA structures. The location of the crossover points can be controlled by strengthening the bonds between their base pairs, the way it was done in the case of a molecule with one junction point [29].

It is necessary to stress that antiparallel DX molecules are rigid, unlike the usual branched structures, which make them suitable for the assembly of nanostructures. In addition, the introduction of the sticky ends in combination with a unique sequence determines the specific association of the rigid building blocks and provides the possibility of controlling it by selecting complementary Watson–Crick sequences. Figure 2.16(A) shows a rigid double-crossover DNA molecule; Figure 2.16(B,C) represent the other variants of double-crossover DNA molecules and their nucleotide sequences. The triangles mark the sites split by enzymes at the biochemical analysis of the structure shown in B. The presence of single-stranded fragments containing five bases (the sticky ends, shown in variant C, I and II) provides the possibility of the hybridization of the two DX molecules. The arrows label the 3′-ends of DNA strands. Both of these parameters are necessary for effective formation of rigid DNA-based nanostructures.

Other rigid structures, such as DNA molecules with triple crossover points, have been theoretically calculated and synthesized (Figure 2.17). A triple crossover (TX) molecule contains three domains, each of them crossing the neighboring domain twice [33].

Figure 2.17(A) demonstrates the reciprocal exchange between hairpins. Here a red hairpin and blue hairpin with zero node between them are presented. Their helix axes are horizontal, and the dyad axis between them is vertical. After reciprocal

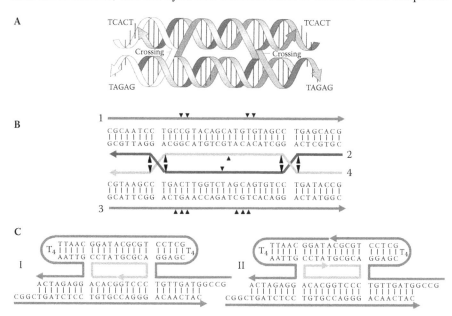

FIGURE 2.16   (**See color insert**) Double-crossover DNA structures with sticky ends used for formation of nanostructures. (Courtesy of Nadrian C. Seeman.)

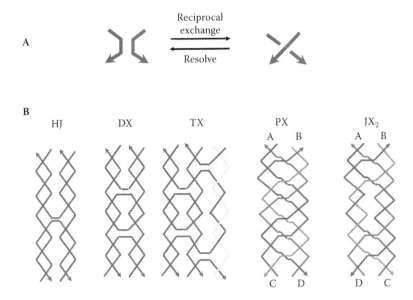

FIGURE 2.17 **(See color insert)** Reciprocal exchange between two DNA hairpins and examples of rigid DNA motifs important in nanotechnology. (Courtesy of Nadrian C. Seeman.)

exchange, the two hairpins have been converted into a single duplex molecule, with the colors indicating that the new molecules are combinations of the old ones. Following this process, one strand is filled-unfilled, and the other strand is unfilled-filled. The reverse operation is called resolution.

Figure 2.17(B) represents the scheme of a Holliday junction (HJ)—a four-arm junction that results from a single reciprocal exchange between double helices. A DX molecule that is formed as the result of a double exchange is shown on the right. Next to the right of the DX molecule is a triple crossover (TX) molecule that results from two successive double reciprocal exchanges. The HJ, DX, and TX molecules all contain exchanges between strands of opposite polarity. To the right of the TX molecule is a paranemic crossover (PX) molecule, where two double helices exchange strands at every possible point where the helices are contacting. Next to the right of the PX molecule is a JX_2 molecule where the two crossovers typical of the PX molecule are absent. The exchanges in the PX and JX_2 molecules take place between the strands with the same polarity.

TX molecules make it possible to form hollow spaces in a structure, which provides the possibility of incorporating in this structure the components that are protruded out of plane of the formed 2-D construction.

Another motif (DX+J) is formed by combining a hairpin with one of the duplex arms of a DX molecule (Figure 2.18, where the hairpin molecule is fused by reciprocal exchange with the central cyclic strand of a DX molecule). The main use of DX+J motifs has been made with the extra DNA domain oriented as nearly as possible normal to the plane defined by the DX helices. This motif was used as a topographic marker to control the forming structures by the AFM [29, 32].

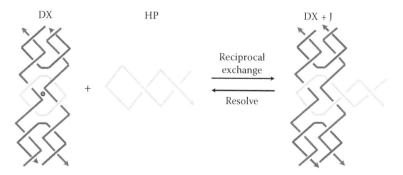

FIGURE 2.18   (See color insert) Reciprocal exchange between a DX molecule and hairpin molecule results in a new motif. (Courtesy of Nadrian C. Seeman.)

Another class of DNA motif is DNA molecules consisting of two parallel double helices that are bound paranemically, so this structure is known as paranemic cross-over DNA or PX DNA (Figure 2.19). (The term *paranemic* means that the melting of this structure leads to the formation of two circular molecules not crossing each other.) PX molecules form junction points at every possible point. This makes the structure more stable in comparison to a DX molecule. The paranemic nature of the PX molecules allows replacement of the sticky ends as long cohesive units.

Figure 2.19 shows two DNA triangles held together by paranemic contacts. One can say that all of the DNA motifs involving fused helices with parallel or antiparallel axes are sufficiently rigid for creation of DNA-based nanostructures.

Molecules containing multiple crossover points are not easy to synthesize and design (in comparison to molecules with one junction point) because two junction points must be phased toward each other. The yield of synthesis is usually about 1%. DNA structures of this kind are mostly analyzed by biochemical methods. For this purpose, nucleotide sequences, split by specific enzymes (endonucleases, see Figure 2.16), are introduced into the structure. This provides the ability to identify

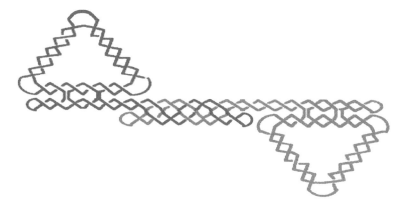

FIGURE 2.19   Paranemic cohesion between DNA triangles.

the parameters of the sugar-phosphate chains in DNA molecules. Electrophoresis is used to determine the integrity of the formed structure.

The DNA molecules containing two or more crossover points are called *DNA tiles* in the literature [34]. DNA tiles can carry single-stranded fragments (sticky ends, see Figure 2.16) that hybridize to sticky ends of other DNA tiles and form complementary complexes. It should be considered that the continuity of sugar-phosphate chains in the helical domains of DX molecules is important for the process. Namely, if the first domain in CI or CII molecules in Figure 2.16 (closed loop consisting of four thymine residues) is highly rigid, the nicked sugar-phosphate chain of the second domain determines its flexibility. Due to the stiffness of CI molecules, the ligase processing results in the formation of relatively long oligomers containing up to seventeen monomers. And, vice versa, in the case of more flexible branched molecules, it is difficult to receive oligomers with a polymerization degree above 6. So, if the tiles are rigid, and the neighboring tiles are coplanar [28], it is possible to create spatial 2-D structures by joining tiles.

Therefore, another type of DNA molecule has been developed—the DNA tiles that are used as building blocks for the hybridization technique of DNA nanostructure formation. Such tiles can be used to hybridize them with other tiles and form flat 2-D nanostructures (Figure 2.20, where A represents a two-component array and B a four-component array). Extra helical domains, such as DX tiles (tiles B^+ and D^+, in which a DNA hairpin protrudes from the plane defined by the helix axes of the two antiparallel domains), can be introduced into the structure and used as topographic markers (shown as black points) visible in the atomic force microscope (AFM). Figure 2.20(A) shows a structure based in two types of tiles (A and B^+). Two helical DX molecules are schematically shown as rectangles with different structures on the ends. These structures represent the sticky ends

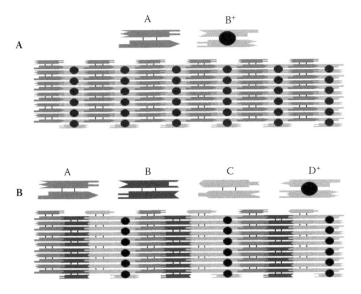

FIGURE 2.20 **(See color insert)** Different variants of flat structures based on DX molecules.

used for hybridization of DX tiles. To assemble DX tiles into planar 2-D super-structures, the total length of arms (including sticky ends) has to correspond to a certain number of half-turns between the crossover points of adjacent tiles. One can see that the black points in the created structure form stripes. Figure 2.20(B) shows a structure consisting of four types of tiles. The designations used in Figure 2.20(A) are retained. As the number and the size of the tiles can be calculated using the model of their assembly, the distance between adjacent stripes evaluated by AFM is used as a criterion to determine the regularity of the model. In the case B, i.e., in the case of a flat structure from four kinds of tiles, the distance must be longer than the distance between stripes in a structure containing two kinds of tiles.

The first 2-D DNA lattices were designed using the DX tiles as well as triple-crossover (TX) complexes [35]. Figure 2.21 (where the size of the observed area is 1400 nm) gives, as an example, a 2-D TX array of DNA nanostructure [36]. The topographic features of the TX+2J molecule appear as stripes in the AFM. The individual strips are separated by 27.2 nm, which corresponds roughly to the expected distance of 28.6 nm.

A schematic image of a PX molecule is shown in Figure 2.19 as an example. This figure shows that two DNA triangles are joined by paranemic contacts [37]. This pattern is also rigid and can be used for the assembly of different nanostructures.

AFM images of some of the structures formed are given in Figure 2.22. The three sets of images correspond to experiments with different magnification: the observation area for (a, d, g) is 1300 nm, that for (b, e, h) is 188 nm, and that for (c, f, i) is maximal magnification. The zigzag character of the strands in (a), (b), and (c) is similar to the expected ligation products. In (c), (f), and (i), the correspondence between the molecular models and the observed images is depicted graphically.

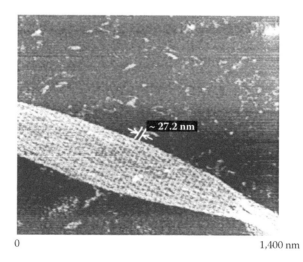

0   1,400 nm

FIGURE 2.21   AFM image of the array of a flat structure formed by DNA tiles.

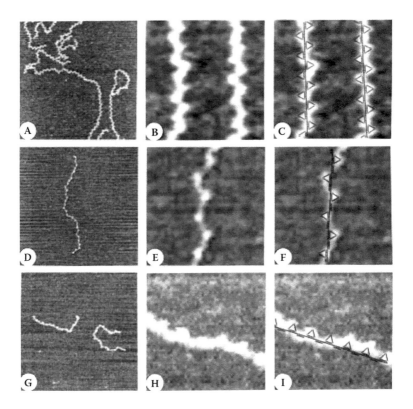

FIGURE 2.22   AFM visualization of triangular ligation products.

Consequently, TX and DX DNA molecules (recombination pattern close to the Holliday structure) can be used as rigid building blocks to form various DNA nano-structures [16].

Recently, a new, highly rigid, three-point-star DNA motif with crossing DNA molecules was used to obtain flat nanostructures with a hexagonal positioning of neighboring elements [38]. It consists of seven DNA single strands organized in three four-arm junctions, as seen in Figure 2.23. In this figure, A depicts a blunt-ended DNA consisting of three red strands, three blue strands, and one dark-green strand; the dark-blue fragments are loops consisting of thymine residues; a black triangle is a threefold rotation axis. Figure 2.23(B) depicts a hexagonal 2-D array assembled with sticky-ended three-point stars. Figure 2.23(C) shows the assembly of a flat struc-ture from three-point-star DNA molecules; the rotation axis between two interacting units is marked with dark arrows. Figure 2.23(D) shows an AFM image of the DNA structure self-assembled from the three-point-star motif with sticky ends; the inset shows the Fourier-transform of the received structure.

The initial structure (A) was formed by hybridization. It consists of seven single-stranded DNA molecules organized in three four-arm junctions. Three T3 loops (dark blue lines) are located at the center to prevent helices from stacking on one another. Junctions are related to each other by threefold rotational symmetry and

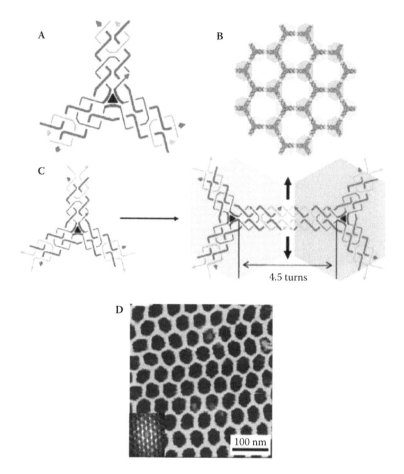

FIGURE 2.23 Formation of a flat structure based on three-point-star DNA motifs.

point from the center to the periphery. The overall shape is a three-point star. The three interconnecting four-arm junctions in this DNA motif constrain each other and force all components to stay in one plane. Because of the threefold symmetry, the seven DNA strands are grouped into three identical red strands, three blue strands, and one dark-green blue strand. When proper sticky ends are added, the three-point-star motif is assembled into an extended, hexagonal 2-D lattice (B). The AFM image (D) of the received array indicates that the period of the lattice is 29.9 ± 0.1 nm, which corresponds to the expected 30.3-nm period calculated from the DNA helical parameters. Though the construction is about 2 nm thick, its size may reach 30 μm. As a result of a vacuum evaporation of gold, a 20-nm-thick metallic replica was obtained, which proves the possibility of using a DNA-based structure as a matrix to receive metallic nanoconstructions. Later, the three-arm structure with crossover DNA molecules was used to obtain a family of symmetric polygons [39].

Clearly, the structures with crossover DNA molecules are an attractive background for the formation of flat extended 2-D nanostructures. The methodology of the self-assembly of various 2-D DNA nanostructures starts with the theoretic calculation and chemical synthesis of single-stranded DNA molecules with specified nucleotide sequences that are joined into tiles for further hybridization. On the whole, the variety of tiles and branched structures combined with simpler elements, such as sticky ends, helices, and loops, makes it possible to create a large number of rigid building blocks for DNA nanostructure production. In most of the examples given here, the initial DNA structures are assembled into flat 2-D nanostructures.

Note that the formation of complicated DNA nanostructures requires multiple reaction stages and refinements of the received intermediate products and building blocks; at the same time, the more complicated the tailored structure is, the lower is the yield of the desired product.

In 2006, a new hybridization technique of nanostructure self-assembly called *single-stranded DNA origami* was introduced. The results of accurate computer calculations of both the structure of the initial DNA molecule and the shape of the final construction was spectacularly displayed by Rothemund [40]. One of the variants of this technique is illustrated in Figure 2.24, where A depicts a hypothetical structure approximated by parallel double helices joined by periodic crossovers, and B shows a finished design after merges and rearrangements along the seam.

The first stage of the technology is the development of a computer model of the final desired nanostructure. Figure 2.24 shows that the designed product is 33 nm wide and 35 nm long. The object is filled with an even number of double helices shown as cylinders. The helices should be synthesized considering the necessary number of turns in each chain. To keep the helices joined together, the synthesized molecules must have single crossovers (blue triangles) where the location of axes of neighboring fragments changes (the way it occurs as a Holliday junction is formed, as seen in Figure 2.12). In such a structure, parallel DNA helices are not densely packed, which is probably caused by the electrostatic repulsion between them. This means that the final size of each dimension is defined by the distance between helices, which depends on the distance between the junction points. Figure 2.24 shows the case where the crossover takes place in every 1.5 turns along the helix, but any

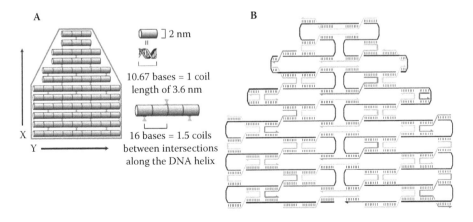

FIGURE 2.24 Formation of flat structure by DNA origami technique.

odd number of turns can be used. In the examined work, the distance between helices was 1 nm when the crossover takes place in every 1.5 turns, and it was 1.5 nm if the transition occurred in every 2.5 turns.

The second stage is the synthesis of a long DNA strand (about 1,000 nucleotides) that is used as the carrying strand (DNA "scaffold" strand). The strand can only fold in certain sites, which leads to the appearance of extra crossover points.

The third stage is a computer calculation and the synthesis of DNA staples, i.e., single-stranded molecules that can form complementary pairs with corresponding sequences in the scaffold strand. The structure and the nucleotide composition of the staples determine the location and the direction of folding of the initial scaffold strand into the final flat structure.

The fourth stage is the calculation of turns in the carrying DNA strand and optimization of their location by adjusting the number of base pairs. It is obvious that at the crossover points of the neighboring staples, a nick (i.e., a break in the DNA backbone) appears in the structure. The nick can be closed by selecting the length of staple molecules. To implement the technology, a natural single-stranded molecule from the M13mp18 virus initially containing 7,249 base pairs and treated with restrictase BsrBI was used as the carrying strand. A 100× surplus (!) of 200–250 corresponding staple molecules was introduced into the solution; the mixture was heated to 95°C and cooled to 20°C within two hours. The cooling was followed by the immobilization of components from the mixture on a film for AFM, and the images were examined by AFM.

As an example, images of some of the received DNA nanostructures are given in Figure 2.25. In this figure, A depicts a triangle with rectangular domains. B shows an acute triangle with trapeziform domains and links between them (red lines on the inset), with the images of structures received by AFM shown in the bottom. The white arrows point to the spots of linking mistakes; the size of the images is 165 × 165 nm; and the scale bar on the images corresponds to 100 nm.

Formation of every nanostructure according to this technology requires prolonged computer modeling followed by one week to synthesize the initial products, although the mixing process, heating, and cooling only takes several hours. The main difficulty is the search of AFM images on a film that fits to the desired assembled nanostructures. Usually the search for one of the desirable structures takes about two days of AFM image analysis.

(To be fair, it should be noted that the application of the ideology based on the specific folding of a synthesized DNA carrying chain [containing 1,669 nucleotides specified by joining five DX staple molecules to the carrier, with each of them containing 40 nucleotides] was first demonstrated in the work of Shin [41]. The hybridization of the obtained mixture led to the formation of an octahedron-shaped nanostructure, although the AFM images of the structure are not very convincing.)

Consequently, the folding of an individual DNA carrying strand into the desirable flat structure controlled by using multiple complementary staple molecules is a new approach to the formation of DNA nanostructures. Unlike the traditional techniques described previously, in this case the hybridization does not require following the strict stoichiometric ratios between the components. The DNA self-assembly occurs

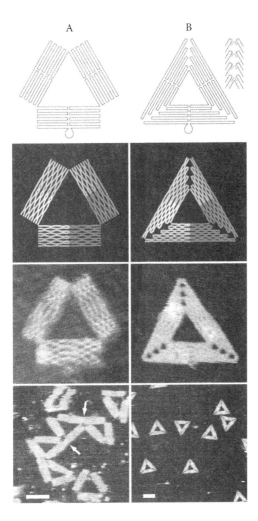

FIGURE 2.25 Examples of flat shapes received with the use of DNA origami technique.

relatively quickly, and the yield of the final product is significantly increased. This is caused by the fact that, after the initial joining of the staple molecules to the carrying strand, the further folding takes place locally within the limits of the carrying strand. Thus the powerful origami technique provides the ability to create DNA nanostructures with a broad variety of shapes, although the process is followed by a large number of assembly mistakes. Hence, the use of a large number of excessive staple molecules is necessary to reduce the number of failures because, in this case, incorrectly joined staples are deleted.

One can say that, in principle, by choosing the desired sequences, one can "program" the interaction between single-stranded DNA molecules and their complementary fragments. In this case, the hybridization technique facilitates the generation of nanostructures (lattices) consisting of alternating rigid ribs and flexible sites. However, to obtain 2-D order in such structures, simple single-stranded components or even Holliday crossover molecules are found to be flexible. This led to the creation of inherently rigid building blocks such as DX or multiple-crossover motifs consisting of DNA duplexes linked together by shared strands. A radically new approach to the creation of DNA nanostructures is based on folding a single-stranded DNA chain (scaffold molecule) into tailored motifs with the help of multiple staple molecules. This scaffold origami technique has certain attractive features that are useful in creating DNA nanostructures.

Thus the hybridization technique based on the application of deliberately constructed single-stranded molecules makes it possible to obtain flat DNA nanostructures of various shapes. Nevertheless, the question of whether the properties of the created nanostructures correspond to the properties typical of the classical nanoparticles (chapter 1) remains open. This is probably due to the low yield of the final products using the considered techniques, which makes the analysis of the properties of these nanostructures with the help of standard physicochemical methods quite difficult or impossible.

So, it is interesting to consider additional methods toward efficient self-assembly of single-stranded DNA molecules.

2.3   ORGANIC LINKER MOLECULES AS RIGID "VERTICES" IN THE TAILORED SPATIAL DNA NANOSTRUCTURES FORMED BY HYBRIDIZATION TECHNIQUE

The junction points necessary for the formation of DNA nanostructures may be both molecules of nucleic acids themselves and molecules of other chemical compounds. A linker (the vertex, the center, the module) can be joined to two or more single-stranded DNA molecules, which can give the system new structural or chemical properties. Even if single-stranded DNA molecules are joined to the linker, their ability to interact and hybridize with complementary nucleic acid molecules is retained, and in this case it becomes possible to form specific spatial structures [31].

Using the strategy proposed by Shi and Bergstrom [42], which can be considered supplementary to the technique proposed by Seeman, building blocks from single-stranded DNA molecules attached to rigid vertices were synthesized. Simple two-arm vertices consisting of two p-(2-hydroxiethyl)phenylethynylphenyl spacers attached to a single tetrahedral carbon atom were used (see Figure 2.26, where the nonhybridized thymidine residues at 5′-ends of the oligonucleotides are shown as small rectangles). The vertices were designed such that oligonucleotide arms are attached through rigid spacers to a single tetrahedral hub.

The structure of the main construction blocks is shown in the inset of Figure 2.26. As a topographic marker, a sequence of three (five) thymidine (T) containing residues was placed at the 5′-ends of single-stranded molecules. Oligonucleotide-conjugated vertices are self-assembled by hybridization into a series of discrete cyclic structures. Depending on the concentration of the building blocks, series of structures containing double-stranded DNA fragments were formed [42]. Such multiarm structures may include from two to six building blocks, and they can be separated by gel electrophoresis under nondenaturing conditions.

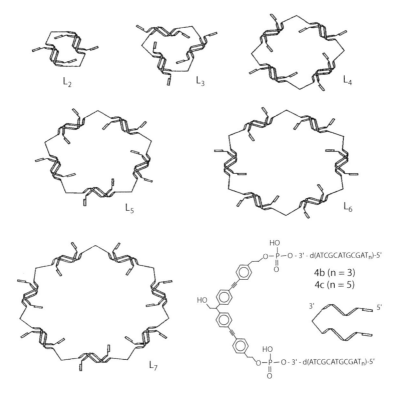

FIGURE 2.26 Self-assembly of DNA cyclic nanostructures based on the use of rigid vertices attached to self-complementary oligonucleotides.

Single-stranded nucleic acids with complementary base sequences joined to an organic linker cannot only hybridize to form double-stranded molecules, but to induce spatial separation of the molecules forming the vertex.

In the work of Gothelf et al. [43], the oligonucleotides were attached to salicylaldehyde groups of two- and three-arm compounds. The final products were called linear oligonucleotide-functionalized modules (LOMs), and thripoidal oligonucleotide-functional modules (TOMs). (Figure 2.27(A) and Figure 2.27(B)) show the chemical structure of these components, respectively.) Modules have sixteen nucleotides at each terminus, which can link up to others containing complementary sequences. The oligonucleotides attached on either side of the salicylaldehyde groups act as clamps to hold the organic modules in a predetermined arrangement. The salicylaldehyde groups of two modules are brought in close proximity when their complementary oligonucleotide sequences are annealed together. When the complementary sequences of the nucleic acids hybridize, salicylaldehyde groups of the two modules approach, i.e., the oligonucleotide molecules regulate the self-assembly. The process of their approaching can be controlled by introducing manganese, Mn, ions that are fixed between the approaching hydroxyl groups of salicylaldehyde molecules (see Figure 2.27(C)).

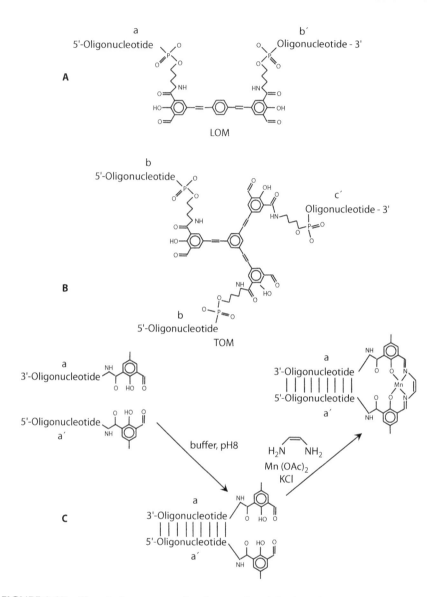

FIGURE 2.27   Chemical structures of various vertices joined to oligonucleotide molecules capable of hybridization.

A set of works on the creation of linkers that can fix the location of three single-stranded DNA molecules in space is of special interest [44–46]. As an example, Figure 2.28 shows the structure of some of these linkers, which can be joined to single-stranded DNA molecules. Figure 2.28(A) shows a branched alkane containing a quaternary carbon atom, and (B) shows the C_{3h} symmetric linker used for fixation of oligonucleotides.

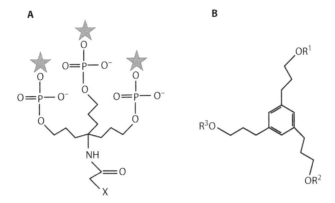

FIGURE 2.28 Chemical linkers capable of joining three oligonucleotide molecules necessary for the formation of spatial nanostructures.

Using twenty such blocks for hybridization, with each block containing single-stranded DNA molecules consisting of fifteen nucleotides, DNA dodecahedra were synthesized [46]. These dodecahedra can carry fluorescent markers, which broadens the possibilities for their practical application.

More complicated organic vertices capable of joining various numbers of DNA molecules can be synthesized to assemble continuous helical structures with various properties. Namely, Ni(II)-1,4,8,11-tetraazacyclotetradecane (nickel cyclam) was used as a linker carrying four identical (or different in nucleotide composition) single-stranded oligonucleotide molecules [47] containing twenty nucleotides each (Figure 2.29). Continuous structures have been formed by introducing complementary single-stranded nucleic acids to various types of Ni cyclam complexes and further hybridization. An example of one of the possible perfect structures is shown in Figure 2.29.

Using the proposed vertex, a periodic lattice with length up to dozens of nanometers can be assembled. Actually, the structure can reach the dimensions of typical macroscopic crystals. Unfortunately, the experimentally formed structure has only been characterized with the help of gel electrophoresis, so it is not clear if the structure assembled from the oligonucleotide (or DNA)-containing building block is linear or branched.

A porphyrin derivate turned out to be an interesting linker (see Figure 2.30, where A shows the structure of a [DNA-porphyrin] conjugate; B shows the sequences of DNA tiles A and B^T; C [top] depicts a 2-D structure obtained from the tiles A and B^T; and C [bottom] shows the 3-D tube formed in the presence of Porph-$(Tc)_4$).

The four ends of tetraporphyrin derivate (Porph-$(Tc)_4$), playing the role of synthetic four-arm linker [48], were joined with four single-stranded molecules of oligonucleotides (ten base pairs). DX tiles (A and B^T) that form a flat 2-D structure under standard hybridization conditions were created deliberately. A supplementary single-stranded molecule containing twelve nucleotides, ten of which were complementary to initial molecules joined to the linker, was attached to the B^T tile. It was supposed that four DX B^T molecules would hybridize with oligonucleotide molecules joined to the linker, and then the molecules of the A tile would hybridize with B^T molecules. In the presence of Porph-$(Tc)_4$, a continuous cylindrical (tube) structure

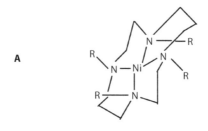

A

R = TCGACTCGACCAGCTCAGCT - O(CH$_2$)$_6$HNCOCH$_2$

B

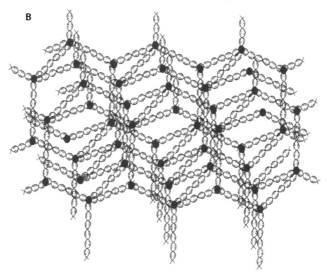

FIGURE 2.29 Chemical linkers based on Ni cyclam with four arms of oligonucleotide molecules (A); hybridization of oligonucleotides may lead to the formation of a layered structure from these molecules (B).

must be formed. The analysis of AFM images indicates the presence of such a tubular structure. Casual inspection of the AFM images shows that the [DNA-porphyrin] complexes are on the outside the tubular structure. However, it is not excluded that they are also on the inside structure, inducing the tube formation.

It is possible that the use of synthetic linkers may simplify the problem of the creation of DNA 3-D nanostructures [49]. In particular, a new method, illustrated in Figure 2.31, was proposed in 2007. In this method, the use of synthetic linkers combined with a set of synthesized DNA-based toolboxes makes it possible to assemble various DNA 3-D nanostructures from a relatively small number of cyclic single-stranded building blocks (Figure 2.31(A)). During the implementation of this method, first the cyclic single-stranded building blocks were synthesized: triangle 3, square 4, pentagon 5, and hexagon 6. This also includes the synthesis of a single continuous DNA strand containing the appropriate number of vertex 1 molecules (for instance, three for structure 3, four for structure 4, etc.) and its subsequent

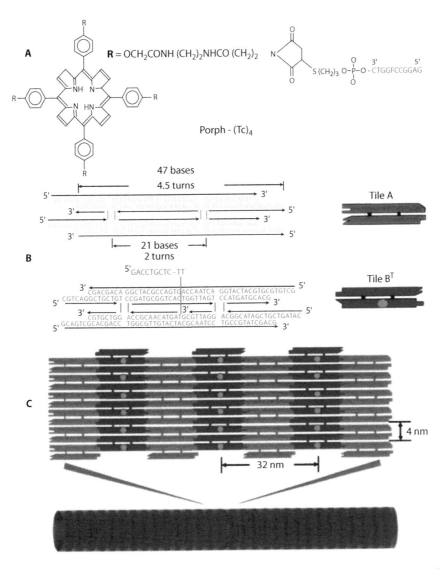

FIGURE 2.30 **(See color insert)** Porphyrin linker and DNA tiles used for the formation of a tubular structure. (Courtesy of Nadrian C. Seeman.)

DNA-templated chemical ligation. The single-stranded and cyclic structure of products 3–6 was confirmed using ExoVII enzymatic treatment. The hybridization of the construction blocks 3–6 has resulted in the formation of 3-D structures. Triangular (P3), rectangular (P4), pentagonal (P5), and hexagonal (P6) structures' prisms were initially assembled. For instance, prism P3 is assembled from two triangular structure-3 units, three linking strands (LS), and three rigid strands (RS) (scheme B). The interesting result of the work is that the created structures can change their spatial shape in response to the effect of exterior factors.

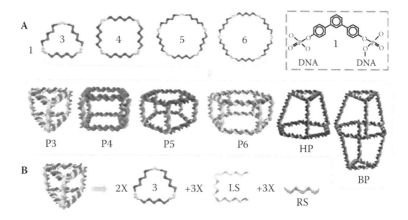

FIGURE 2.31   Building blocks used for the creation of DNA 3-D structures.

The application of synthetic organic linkers (vertices) makes it possible to create DNA nanostructures with various spatial shapes. However, the question of whether the properties of the created nanostructures correspond to the properties of classical nanoparticles has not been analyzed in the previously cited works. As noted earlier, this is because of the complicated technique of preparation of DNA nanostructures and the very low yield of the final product.

In this respect, the work of Dorenbeck [50] should be highlighted because it efficiently uses the concept that three single-stranded DNA molecules can be fixed around the vertex so that their mutual location is pyramidal. As the result of a targeted synthesis, vertices have been obtained, and single-stranded DNA molecules containing 9, 15, and 20 bases were attached to them. The hybridization of these building blocks has resulted in the appearance of the tetrahedrons shown in Figures 2.32(A) and 2.32(B), where A depicts the tetrahedron with ribs containing nine nucleotide pairs and B shows the tetrahedron with ribs of fifteen nucleotide pairs. The gray spheres are the vertices linking three oligonucleotides.

The obtained tetrahedrons have been described using various physicochemical methods including AFM, although the AFM images, due to the limited sensitivity of the method to minor objects, do not provide an exact structure of the received constructions.

The most significant is the fact that the melting curves of the tetrahedrons, each rib of which consisted of nine complementary base pairs and fifteen complementary base pairs, have been obtained for the first time in this work. The comparison of the melting curves (Figure 2.32(C)) shows that a tetrahedron with a minor number of base pairs is characterized by an exponential melting curve (1), and the melting temperature (T_m) value is close to 25°C. At the same time, for a tetrahedron with the ribs consisting of fifteen base pairs, the melting curve is S-shaped (2), and the T_m reaches 50°C. Processing the obtained curves according to the Gibbs–Thomson equation (chapter 1), and then comparing these data to the results given before that describe the melting of nanostructures formed by double-stranded oligonucleotides (Figure 2.33), testifies to the close correlation between the results of different authors.

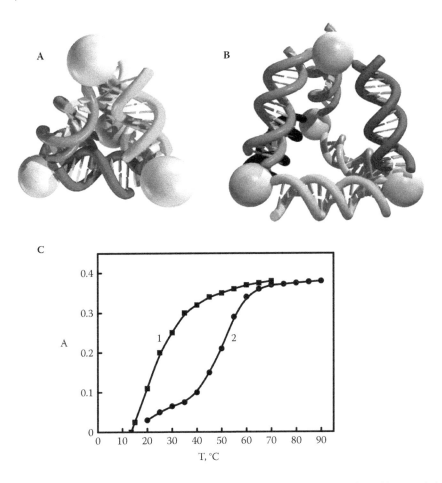

FIGURE 2.32 The structures of nanometric tetrahedrons consisting of double-stranded DNA ribs, and the melting curves of the nanometric tetrahedrons with ribs containing nine (curve 1) and fifteen (curve 2) nucleotide pairs.

The result reported by Dorenbeck [50] corresponds to the data on the melting of nanostructures formed from double-stranded nucleic acids given in chapter 1 and has the most significant, fundamental meaning. This result makes it possible to state the following: The nanostructures formed by materials based on nucleic acid components can be characterized by the same features that reflect the similar peculiarities (not only the size) of the known nanostructures.

2.4   ARRANGEMENT OF OBJECTS ON THE SURFACE OF DNA NANOSTRUCTURES

The achievements of the hybridization technique for creating DNA nanostructures with retained chemical reactivity of all structural elements of DNA have stimulated attempts to apply nanostructures as matrices capable of regularly locating chemically

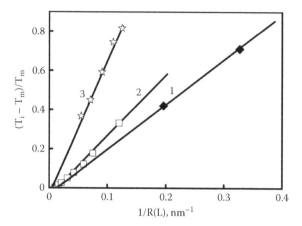

FIGURE 2.33 Size-dependence of the relative melting temperature for nanometric tetrahedrons from double-stranded oligonucleotides (curve 1, black rhombs), Sn-nanoparticles (curve 2, open squares), and double-stranded oligo-(AU) (curve 3, stars).

or biologically relevant compounds on the surface of these matrices [28, 31, 51]. The number of works in this area is relatively small, so several of the most significant approaches can be summarized here.

Four-arm structures with DNA tiles have been formed by hybridizing nine single-stranded DNA molecules [52]. See Figure 2.34, where A is a 4 × 4 DNA tile used as a building block (such tiles contain nine oligonucleotides shown as simplified backbone traces). One four-arm junction is oriented in each direction (N, S, E, W), and the red strand participates in all four junctions and contains loops from four thymine

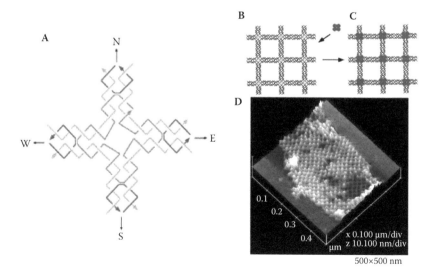

FIGURE 2.34 Self-assembly of 4 × 4 DNA tile, fixation of streptavidin molecules on a nanogrid consisting of DNA tiles, and AFM image of the ordered protein molecules.

residues (T_4) connecting neighboring junctions. Figures 2.34(B, C) are schematic drawing of the DNA grid, which assembles biotin molecules incorporated into one of the loops at the center of each tile. Figure 2.34(D) is an AFM image of the self-assembled protein molecules, with the size scale shown below the figure.

The sequences of nitrogen bases in the initial single-stranded DNA molecules have been selected so that, at the hybridization loops consisting of noncomplementary thymine, residues would appear in the center of each tile. Regardless of the presence of a loop in the central part of the given structure, the crossover of DNA molecules in every arm makes the tiles rigid enough. After the formation of sticky ends in the tiles, they self-assemble into flat nanostructures, whose properties can be controlled by AFM. The analysis indicates that the created nanostructures contain square hollows.

To fix the biologically important compound—the protein streptavidin—on the surface of the created structure, the following technique was used. Chemical groups of thymine residues in the central loop of the tiles were chemically modified so that they keep a biotin molecule, and a flat nanostructure was formed from the modified particles.

Then streptavidin, which forms a stable complex with biotin, was introduced into the construction. Streptavidin molecules have also been fixed on the surface of one-dimensional and two-dimensional structures formed by hybridization of two types of various four-arm structures containing crossing single-stranded DNA molecules [53]. The selection of the corresponding base sequences for the formation of the initial single-stranded DNA molecules was realized by the SEQUIN computer program. The selection of sticky ends in the tiles resulted in self-induced hybridization and formation of two- or one-dimensional flat continuous nanostructures. The introduction of streptavidin into the formed two- or one-dimensional structures was followed by its fixation on the surface of the structures.

As the size of streptavidin is about 5 nm, its interaction with biotin bonded to the DNA nanostructure leads to the appearance of regularly located spherical structures on the surface of the formed grid. These objects can be used as topographic markers in the formed grid, as they are easily detected with AFM (Figure 2.34(D)). In particular, the AFM image testifies to the regular location of streptavidin molecules on the 2-D structure formed from two types of tiles. The distance between adjacent streptavidin molecules is 36.9 nm in the case where streptavidin is bonded to biotin molecules included into one type of tiles, which corresponded to the calculated parameters. One-dimensional DNA nanostructures looking like a railroad bed, with streptavidin molecules fixed at a regular distance from each other, have also been formed from these tiles.

Two-dimensional DNA structures make it possible to order metal nanoparticles, namely, gold nanoparticles [54]. Four construction DNA tiles—A, B, C, D (see Figure 2.35, where A-D-T is the sequence of nitrogen bases in different DX tiles and their geometric arrangement)—have been formed from a number of synthetic oligonucleotides (twenty-one nitrogen bases in each). The interactions between these tiles are programmed by the design of sticky ends. A supplementary single-stranded DNA molecule that could bond metal nanoparticles was introduced into tile B. Tile D contained two loops that served as topographic markers. A gold nanoparticle

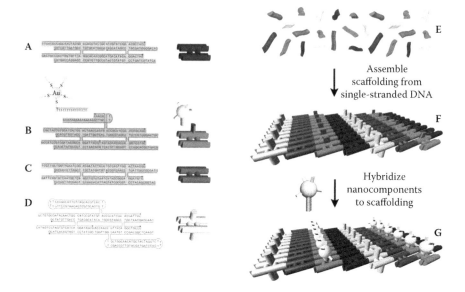

FIGURE 2.35   **(See color insert)** Base sequence and the structure of DX tiles (a unique color corresponds to each DX tile) and assembly steps for 2-D gold nanoparticles array.

(6 nm) was functionalized with (dT)$_{15}$-oligonucleotide strands with a thiol group on the 3′-end. (The fifteen thymine residues were complementary to the sequence from fifteen adenine residues in the B DX tile.) The fixation of gold nanoparticles on the surface of the 2-D DNA nanostructure was conducted by following a few steps.

At first (Figure 2.35), 2-D tile was assembled from single-stranded DNA molecules by the standard hybridization technique. A drop of solution containing the 2-D DNA nanostructure was placed on a surface of freshly split mica. Then the surface of the grid was treated with solution containing the conjugate (gold-(dT)$_{15}$ nanoparticle) to start the hybridization process. The hybridization of the tiles is determined by the presence of sticky ends. One spherical gold particle bears 5–10 identical strands of a single-stranded oligonucleotide that are complementary to the hybridization on tile B. Tile B (red) has a single-stranded oligonucleotide able to hybridize further. Tile D contains topological markers (yellow) protruding from its spatial structure. Different colors are used to illustrate the peculiarities of the forming structure. The right-hand side of Figure 2.35 (E–G) represents the steps of the 2-D nanostructure assembly. The scaffold is processed with oligonucleotide-containing particles, which leads to the hybridization and fixation of the particles on the grid surface. The scheme shows that each gold nanoparticle is only bonded to one site on the scaffold.

Figure 2.36 gives AFM images of gold nanoparticles on the surface of a DNA lattice (A) and an image received by electron microscopy (B). Particles with high contrast have a size of 6.2 ± 0.8 nm. The center-to-center distance between the particles in a row varies from 15 to 25 nm; in closely packed areas the distance is 12.0 ± 3.0 nm. The average spacing between rows is 62.9 ± 0.8 nm. The proposed method can be optimized by more thorough selection of the conditions of the DNA chain hybridization and the methods of metal nanoparticle functionalizing.

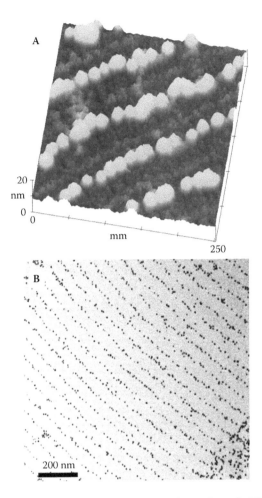

FIGURE 2.36   Visualization of gold nanoparticles on the surface of a DNA lattice.

In the work of Pinto et al. [55], a very similar technology was used to form more complex nanostructures. Four types of DX DNA molecules (A, B, C, D) were again used. The tiles are different by the presence of supplementary single-stranded molecules introduced into tiles B and D (Figure 2.37).

In addition, tile D contained a loop that was used as a topographic marker. The presence of sticky ends in the tiles caused self-assembly of the tiles into a flat nanostructure so that two tiles, A and B, determined the distance of 32 nm between tiles C and D in the received structure. Gold nanoparticles with a diameter of 5 and 10 nm were bonded to single-stranded $(dT)_{15}$ or $(dT)_7$ molecule complementary to one of the single-stranded DNA molecules in tiles B and D. Such nanoparticles were used as objects placed on the surface of a 2-D structure. The left-hand side of Figure 2.37(A–D) shows the base sequence in different tiles (DX molecules). Hybridization of the tiles is determined by the presence of the sticky ends. Each spherical gold particle bears a shell composed of multiple, identical DNA strands (5–10 strands per 5 nm

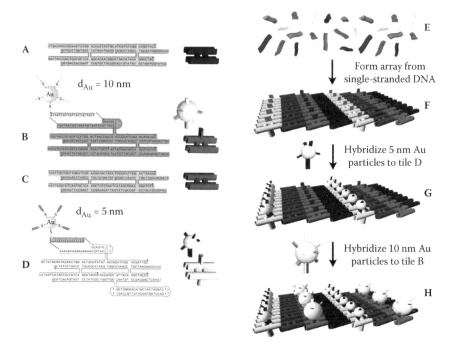

FIGURE 2.37 **(See color insert)** Base sequence and spatial structure of DNA tiles used to form a lattice containing gold nanoparticles of different size and the assembly steps.

particle and 60–70 per 10 nm particle) that are complementary to the hybridization sites on tiles B and D. Sequences from $(dT)_{15}$ and $(dT)_7$ were used in the case of tile D; Figure 2.37 shows the $(dT)_{15}$ variant. Different colors are used to illustrate the peculiarities of the structure. The right-hand side of Figure 2.37 shows the steps in assembling the 2-D structure.

After forming a flat structure from four tiles, the received structure (G) was treated with a solution containing 5-nm gold particles that, due to the single-stranded oligonucleotides bonded to them, hybridize with an open hybridization spot on a D tile. The grid is processed with other gold nanoparticles (H) that, due to the single-stranded oligonucleotides bonded to them, hybridize to an open hybridization site on a B tile. The AFM image in Figure 2.38 indicates that the nanoparticles look like parallel stripes with a distance of 64 nm between them, and the average distance between adjacent gold nanoparticles is about 10 nm.

A slightly different approach to the ordering of gold nanoparticles was described in work of Zheng et al. [56]. A triangle formed from DX tiles (Figure 2.39) was used as the initial element for the formation of a 2-D DNA nanostructure. (In Figure 2.39, A is a hypothetical 2-D structure from DNA DX tiles containing 5- and 10-nm gold nanoparticles; B is an AFM image of a 2-D structure from DNA DX tiles.)

Gold nanoparticles (5 or 10 nm) were attached to one or two different triangles. The triangles hybridize and form a rhomboid grid. The ACM image gives a concept of the structure of the assembled 2-D grid (Figure 2.39(B)), and the electron

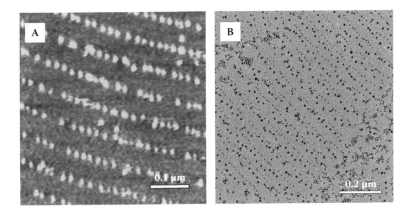

FIGURE 2.38   AFM image of a DNA nanolattice with various gold nanoparticles (A); and (B) electron microscopy image of a DNA grid containing gold nanoparticles with different sizes; rows of parallel stripes of major and minor gold particles can be seen.

microscopic image shows clearly that the gold nanoparticles can be ordered on the surface of such DNA structures.

The examples given here indicate that the use of rigid DNA tiles synthesized in advance makes it possible to both create DNA 2-D nanostructures and controllably locate gold nanoparticles on the surface of such structures. It is probable that such DNA-based objects can be applied in technology. Basically, several approaches proposed in the literature [54–56] can provide the self-assembly of nanocomponents into objects of microscopic size.

Consequently, on the one hand, the hybridization technique provides the possibility of creating DNA 2-D nanostructures and then fixing chemically or biologically

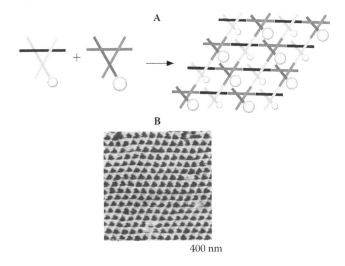

FIGURE 2.39   The 2-D structure assembled from DNA DX tiles bonded to gold nanoparticles of different sizes.

relevant compounds on their surfaces. On the other hand, the number of such compounds may be very limited, as the compounds themselves must not alter the spatial structure of nitrogen bases in the initial DNA molecules used to form the rigid tiles, and they can retain their chemical structure during all steps of the hybridization self-assembly of the DNA lattices. Finally, the questions regarding the physicochemical properties of the created structures, the yield of the targeted product, and the total cost of the technology remain open.

Currently, the hybridization technique cannot provide the formation of continuous DNA 3-D nanostructures and their complexes. There is no answer to the question of whether a high concentration of guest molecules—chemically or biologically relevant compounds important for researchers—can be retained in tailored nanostructures. Indeed, summarizing the results of different variants of the hybridization technique of DNA nanostructure assembly as a hypothetical scheme shown in Figure 2.40, one can evaluate the ability of such structures to carry a high concentration of guest molecules. According to the results given in this chapter, guest molecules can be accumulated in a nanostructure using the free space available in the DNA secondary structure. A DNA molecule forming rigid ribs in the structure can carry such compounds as intercalators (i.e., planar compounds that intercalate between pairs of nitrogen bases) or the compounds that locate in one of the grooves on the surface of this macromolecule (polypeptides, proteins, etc.). The equilibrium concentration of any of the carried compounds depends on the peculiarities of the DNA secondary structure (distance between base pairs, their inclination, etc., as well as charge density of phosphate groups, etc.) and obeys the known laws of absorption. Hence, the maximal number of guest molecules depends on the peculiarities of the DNA secondary structure and, consequently, is limited. This number, under all identical conditions, also depends on the structure of the used guest molecule.

It is obvious from the physicochemical properties of the initial DNA molecules that two approaches to accumulation of the guest molecules in a DNA nanostructure

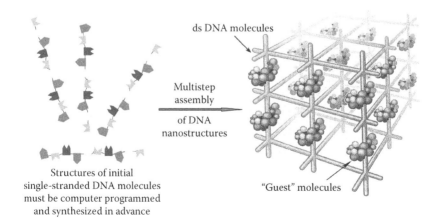

FIGURE 2.40   (See color insert) Hypothetical structure formed by hybridization technique from single-stranded DNA molecules that can carry "guest" molecules.

are possible. According to the first one, the guest molecules can be bonded to the initial single-stranded DNA molecule and, considering that the properties of the DNA molecule are constant, the necessary 3-D structure can be formed from a [DNA-guest] complex using the hybridization technique. At the same time, it is obvious that the presence of several biochemical and physicochemical steps in the process of the formation of the nanostructure realized under different conditions influence the probability of retaining the high concentration of the guest molecules. So, the second approach seems to be more probable. According to the second approach, the necessary 3-D structure can be formed from the initial single-stranded DNA molecules by hybridization, and then the guest molecules can be bonded to double-stranded DNA fragments forming the rigid ribs in the formed nanostructure. Even if the physicochemical properties of the DNA molecules in this structure are not changed during hybridization and assembly processes, these considerations dictate that the number of guest molecules carried by DNA molecules cannot be significantly different from the number of molecules that every initial, individual double-stranded DNA molecule can carry in its secondary structure. So, by this approach, the probability of retaining a high concentration of guest molecules in the structure shown in Figure 2.40 is still unlikely.

2.5 SUMMARY

The data presented in this chapter prove that a single-stranded DNA molecule is the most attractive object as a building block in nanobiotechnology. The hybridization technique facilitates the creation of various 2-D nanostructures from synthetic single-stranded DNA molecules. The key advantage of the application of such DNA molecules as building blocks is the opportunity to realize the specific recognition between the complementary molecules based on the presence of the sticky ends. Various additional DNA building blocks such as DNA cruciform structure, the Holliday junctions, double- or multiple-crossover DNA motifs, various DNA tails, as well as DNA scaffold and DNA staples used in an origami technique, etc., were used to obtain nanoobjects. Hence, by properly choosing a defined sequence, one can program the process of hybridization of single-stranded DNA chains and use them to generate planar nanolattices. The peculiarities of the different variants of hybridization methods are also considered and illustrated in this chapter.

The advantage of single-stranded DNA as a building block is strengthened by the possibility of automatic synthesis of any DNA sequence and the possibility of amplifying any sequence from microscopic to macroscopic amounts due to the polymerase chain reaction. However, automated DNA synthesis can only prepare sequences of up to about one hundred bases. To build larger structures, biological techniques must be developed that either splice together short artificial DNA strands or allow the synthesis of longer strands. Such methods exist for natural DNA, but these would have to be modified to tolerate artificial bases.

An important advantage of a DNA molecule used as a construction material is its relatively high physicochemical stability, which significantly exceeds the stability of other biopolymers, such as proteins. This means that nanomaterials based on DNA molecules can be kept at a wide range of experimental conditions without special safety measures applied to avoid the decay of a biological material. The latter is distinctively illustrated by the analysis of the DNA sequence from the bones of a Neanderthal skeleton [57], verifying that the molecules were retained even after 50,000 years. The enumerated advantages of DNA have already been used to assemble nanometric objects. The sharp increase in the number of publications in this area shows that the creation of DNA-based nanostructures via the hybridization technique is more than just an exciting intellectual exercise.

REFERENCES

1. Niemeyer, C. M. 2001. Nanoparticles, proteins, and nucleic acids: Biotechnology meets materials science. *Angew. Chem. Int. Ed.* 40:4128–58.
2. Yan, H., T. H. LaBean, L. P. Feng, and J. H. Reif. 2003. Directed nucleation assembly of DNA tile complexes for barcode-patterned lattices. *Proc. Natl. Acad. Sci. USA* 100:8103–8.
3. Yevdokimov, Yu. M., and V. V. Sytchev. 2008. Principles for creating nanoconstructions with nucleic acid molecules as building blocks. *Uspekhi Khimii* (Russian ed.) 77:194–206.
4. Seeman, N. C. 1982. Nucleic acid junctions and lattices. *J. Theor. Biol.* 99:237–47.
5. Seeman, N. C. 2007. An overview of structural DNA nanotechnology. *J. Mol. Biotechnol.* 37:246–57.
6. Sun, Y., and C. H. Kiang. 2005. DNA-based artificial nanostructures: Fabrication, properties, and applications. In *Handbook of Nanostructured Biomaterials and Their Applications in Nanobiotechnology*, vols. 1–2, ed. H. S. Nalwa, chap. 5. Los Angeles: American Scientific Publishers.
7. Niemeyer, C. M. 1999. Progress in "engineering up" nanotechnology devices utilizing DNA as a construction material. *Appl. Phys. A* 68:119–24.
8. Demidov, V. V., and M. D. Frank-Kamenetskii. 2004. Two sides of the coin: Affinity and specificity of nucleic acid interactions. *Trends Biochem. Sci.* 29:62–71.
9. Pool, Ch. P., and F. J. Owens. 2003. *Introduction to Nanotechnology.* New York: John Wiley & Sons.
10. Saenger, W. 1984. *Principles of Nucleic Acid Structure.* New York: Springer-Verlag.
11. Lehninger, A. L., D. L. Nelson, and M. M. Cox. 1993. *Principles of biochemistry.* 2nd ed. New York: Worth Publishers.
12. Seeman, N. C. 2000. In the nick of space: Generalized nucleic acid complementarity and DNA nanotechnology. *Synlett.* 10:1536–48.
13. Seeman, N. C. 1985. Macromolecular design, nucleic acid junctions and crystal formation. *Biomol. Struct. Dyn.* 3:11–34.
14. Seeman, N. C. 1997. DNA components for molecular architecture. *Acc. Chem. Res.* 30:357–63.
15. Seeman, N. C. 2005. DNA enables nanoscale control of the structure of matter. *Quart. Rev. Biophys.* 38:363–71.
16. Seeman, N. C. 1999. DNA engineering and its application to nanotechnology. *Trends Biotech.* 17:437–43.
17. Seeman, N. C. 1990. De novo design of sequences for nucleic acid structure engineering. *J. Biomol. Struct. Dyn.* 8:573–81.
18. Brenneman, A., and A. E. Condon. 2002. Strand design for biomolecular computation. *J. Theor. Comput. Sci.* 287:39–58.

19. Kallenbach, N. R., R.-I. Ma, and N. C. Seeman. 1983. An immobile nucleic acid junction constructed from oligonucleotides. *Nature* 305:829–31.

20. Seeman, N. C., and N. R. Kallenbach. 1983. Design of immobile nucleic acid junction. *Biophys. J.* 44:201–9.

21. Seeman, N. C. 2004. Nanotechnology and the double helix. *Sci. Am.* 290:64–69.

22. Holliday, R. 1964. A mechanism for gene conversion in fungi. *Genet. Res. Camb.* 5:282–304.

23. Li, X., H. Wang, and N. C. Seeman. 1997. Direct evidence for Holliday junction crossover isomerization. *Biochemistry* 36:4240–47.

24. Lu, M., Q. Guo, N. C. Seeman, and N. R. Kallenbech. 1991. Parallel and antiparallel Holliday junction differ in structure and stability. *J. Mol. Biol.* 221:1419–32.

25. Sha, R., F. Liu, and N. C. Seeman. 2000. Direct evidence for spontaneous branch migration in antiparallel Holliday junction. *Biochemistry* 39:11514–22.

26. Murchle, A. I. H., R. M. Clegg, E. von Kitzing, D. R. Duckett, S. Diekmann, and D. M. J. Lilley. 1989. Fluorescence energy transfer shows that four-way DNA junction is a right-handed cross of antiparallel molecules. *Nature* 341:763–66.

27. Mao, C., W. Sun, and N. C. Seeman. 1999. Designed two-dimensional DNA Holliday junction array visualized by atomic force microscopy. *J. Amer. Chem. Soc.* 121:5437–43.

28. Condon, A. 2006. Designed DNA molecules: Principles and application of molecular nanotechnology. *Nat. Rev. Genet.* 7:565–75.

29. Fu, T. J., and N. C. Seeman. 1993. DNA-double-crossover molecules. *Biochemistry* 32:3211–20.

30. Li, X., X. Yang, J. Qi, and N. C. Seeman. 1996. Antiparallel DNA double crossover molecules as components for nanoconstruction. *J. Amer. Chem. Soc.* 118:6131–49.

31. Feldkamp, U., and C. M. Niemeyer. 2006. Rational design of DNA nanoarchitectures. *Angew. Chem. Int. Ed.* 45:1856–76.

32. Seeman, N. C., H. Wang, X. Yang, et al. 1998. New motif in DNA nanotechnology. *Nanotechnology* 9:257–73.

33. Seeman, N. C. 2001. DNA nicks and nodes and nanotechnology. *Nano Lett.* 1:22–26.

34. Yan, H., T. H. LaBean, L. P. Feng, and J. H. Reif. 2003. Directed nucleation assembly of DNA tile complexes for barcode patterned lattices. *Proc. Natl. Acad. Sci. USA* 100:8103–8.

35. Winfree, E., F. Liu, L. A. Wenzler, and N. C. Seeman. 1998. Design and self-assembly of two-dimensional DNA crystals. *Nature* 394:539–44.

36. LaBean, T., H. Yan, J. Kopatsch, et al. 2000. The construction, analysis, ligation and self-assembly of DNA triple crossover complexes. *J. Amer. Chem. Soc.* 122:1848–60.

37. Yang, X., L. A. Wenzler, J. Qi, X. Li, and N. C. Seeman. 1998. Ligation of DNA triangles containing double crossover molecules. *J. Amer. Chem. Soc.* 120:9779–86.

38. He, Y., Y. Chen, H. Liu, A. E. Ribbe, and C. Mao. 2005. Self-assembly of hexagonal DNA two-dimensional (2-D) arrays. *J. Amer. Chem. Soc.* 127:12202–3.

39. He, Y., T. Ye, M. Su, et al. 2008. Hierarchical self-assembly of DNA into symmetric supramolecular polyhedra. *Nature* 452:198–201.

40. Rothemund, P. W. 2006. Folding DNA to create nanoscale shapes and patterns. *Nature* 440:297–302.

41. Shin, W. M., J. D. Quispe, and G. F. Joyce. 2004. A 1.7-kilobase single-stranded DNA that folds into nanoscale octahedron. *Nature* 427:618–21.

42. Shi, J., and D. E. Bergstrom. 1997. Assembly of novel DNA cycles with rigid tetrahedral linkers. *Angew. Chem. Int. Ed.* 36:111–13.

43. Gothelf, K. V., A. Thomsen, M. Nielsen, E. Clo, and R. S. Brown. 2004. Modular DNA-programmed assembly of linear and branched conjugated nanostructures. *J. Amer. Chem. Soc.* 126:1044–46.

44. Scheffler, M., A. Dorenbeck, S. Jordan, M. Wustefeld, and G. von Kiedrowski. 1999. Self-assembly of trisoligonucleotides: The case for nano-acetylene and nano-cyclobuta-diene. *Angew. Chem. Int. Ed.* 38:3312–15.
45. Eckardt, L. H., K. Naumann, W. M. Pankau, et al. 2002. Chemical coping of connectivity. *Nature* 420:286.
46. Zimmermann, J., M. P. J. Cebulla, S. Monningshoff, and G. von Kiedrovski. 2008. Self-assembly of a DNA dodecahedron from 20 trisoligonucleotides with C_{3h} linkers. *Angew. Chem. Int. Ed.* 47:3626–30.
47. Stewart, K. M., and L. W. McLaughlin. 2004. Four-arm oligonucleotide Ni(II)-cyclam-centered complexes as precursors for the generation of supramolecular periodic assemblies. *J. Amer. Chem. Soc.* 126:2050–57.
48. Endo, M., N. C. Seeman, and T. Majima. 2005. DNA tube structures controlled by a four-way-branched DNA connector. *Angew. Chem. Int. Ed.* 44:6074–77.
49. Aldaye, F. A., and H. F. Sleiman. 2007. Modular access to structurally switchable 3-D discrete DNA assemblies. *J. Amer. Chem. Soc.* 129:13376–77.
50. Dorenbeck, A. 2000. DNA-Nanostrukturen durch self-assembly von trisoligonucleo-tidylen. Dissertation zur Erlangung des Doktorgrades der Naturwissenschaften an der Fakultät Chemie der Ruhr-Universität Bochum. Bochum.
51. Katz, E., and I. Willner. 2004. Integrated nanoparticle-biomolecular hybrid systems: Synthesis, properties and applications. *Angew. Chem. Int. Ed.* 43:6042–6108.
52. Yan, H., S. H. Park, G. Finkelstein, J. H. Reif, and T. H. LaBean. 2003. DNA-templated self-assembly of protein arrays and highly conductive nanowires. *Science* 301:1882–84.
53. Park, S. H., P. Yin, Y. Liu, J. H. Reif, T. H. LaBean, and H. Yan. 2005. Programmable DNA self-assemblies for nanoscale organization of ligands and protein. *Nano Lett.* 5:729–33.
54. Le, J. D., Y. Pinto, N. C. Seeman, K. Musier-Forsyth, T. A. Taton, and R. A. Kiehl. 2004. DNA-templated self-assembly of metallic nanocomponent arrays on a surface. *Nano Lett.* 4:2343–47.
55. Pinto, Y. Y., J. D. Le, N. C. Seeman, K. Musier-Forsyth, T. A. Taton, and R. A. Kiehl. 2005. Sequence-encoded self-assembly of multiple-nanocomponent arrays by 2-D DNA scaffolding. *Nano Lett.* 5:2399–2402.
56. Zheng, J. W., P. E. Constantinou, C. Micheel, A. P. Alivisatos, R. A. Kiehl, and N. C. Seeman. 2006. Two-dimensional nanoparticle arrays show the organizational power of robust DNA motifs. *Nano Lett.* 6:1502–4.
57. Krings, M., A. Stone, R. W. Schmitz, H. Krainitzki, M. Stoneking, and S. Paabo. 1997. Neandertal DNA sequences and the origin of modern humans. *Cell* 90:19–30.

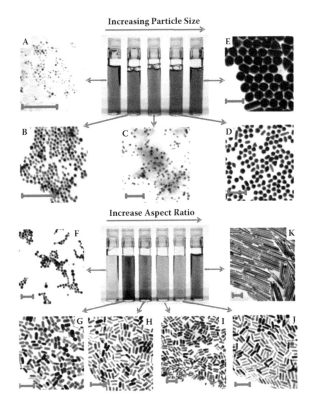

FIGURE 1.1 Change in color of water solutions of spherical gold nanoparticles (upper panels) and gold nanorods (lower panels), depending on the dimensions of the particles.

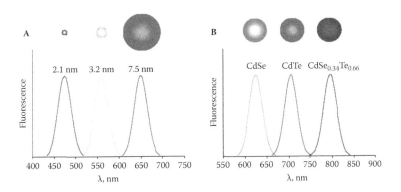

FIGURE 1.7 Some properties of quantum dots with various size and composition: three CdSe quantum dots with different diameter (A) and three quantum dots with different composition (average diameter is about 5 nm) (B) and their fluorescence spectra.

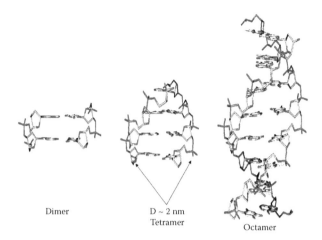

Dimer

D ~ 2 nm
Tetramer

Octamer

FIGURE 1.10 Hypothetical scheme of the formation of a helical structure from nucleotide molecules. (Kindly provided by S. Basili, Padova University, Italy.)

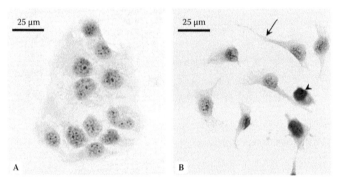

FIGURE 1.16 Cytopathologic analysis of human tumor cells (line H596).

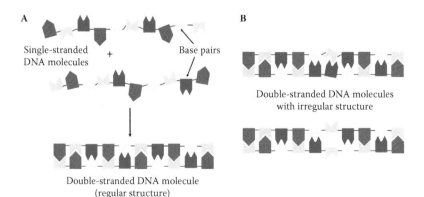

FIGURE 2.8 Formation of a double-stranded DNA molecule as a result of the complementary recognition mechanism of nucleation-zipping.

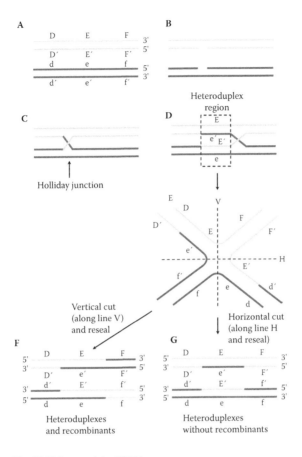

FIGURE 2.12 The Holliday model of DNA crossover.

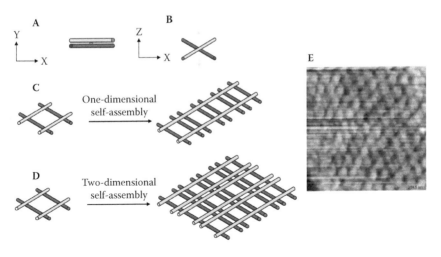

FIGURE 2.14 Lattices built from branched junctions. (Courtesy of Nadrian C. Seeman.)

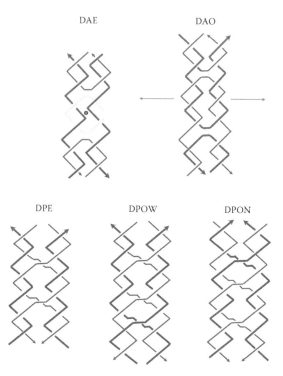

FIGURE 2.15 Five various types of DNA double-crossover molecules. (Courtesy of Nadrian C. Seeman.)

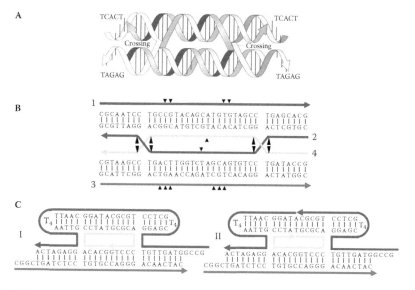

FIGURE 2.16 Double-crossover DNA structures with sticky ends used for formation of nanostructures. (Courtesy of Nadrian C. Seeman.)

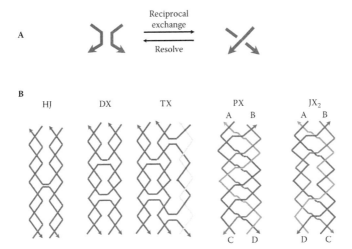

FIGURE 2.17 Reciprocal exchange between two DNA hairpins and examples of rigid DNA motifs important in nanotechnology. (Courtesy of Nadrian C. Seeman.)

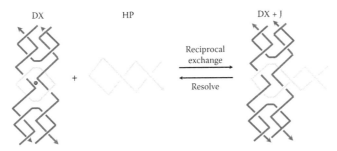

FIGURE 2.18 Reciprocal exchange between a DX molecule and hairpin molecule results in a new motif. (Courtesy of Nadrian C. Seeman.)

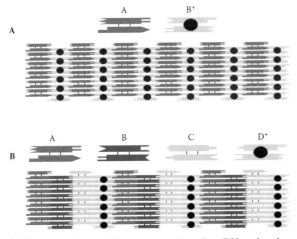

FIGURE 2.20 Different variants of flat structures based on DX molecules.

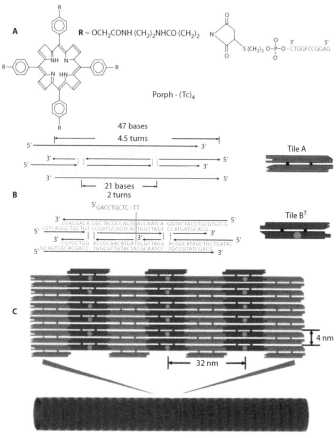

FIGURE 2.30 Porphyrin linker and DNA tiles used for the formation of a tubular structure. (Courtesy of Nadrian C. Seeman.)

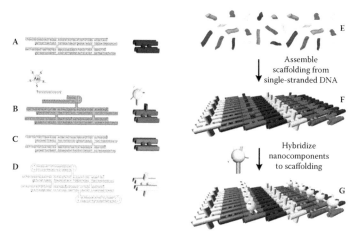

FIGURE 2.35 Base sequence and the structure of DX tiles (a unique color corresponds to each DX tile) and assembly steps for 2-D gold nanoparticles array.

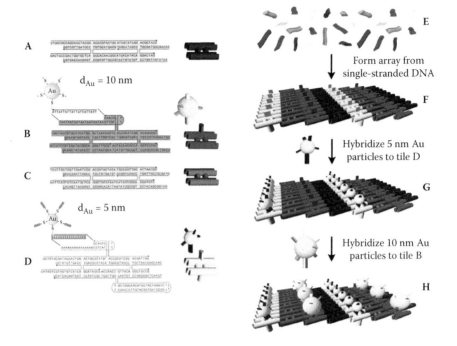

FIGURE 2.37 Base sequence and spatial structure of DNA tiles used to from a lattice containing gold nanoparticles of different size and the assembly steps.

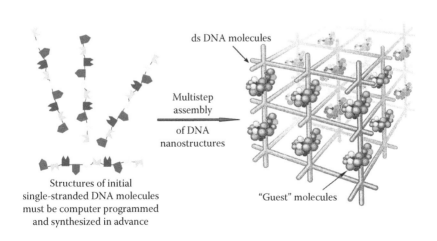

Structures of initial
single-stranded DNA molecules
must be computer programmed
and synthesized in advance

ds DNA molecules

Multistep
assembly
of DNA
nanostructures

"Guest" molecules

FIGURE 2.40 Hypothetical structure formed by hybridization technique from single-stranded DNA molecules that can carry "guest" molecules.

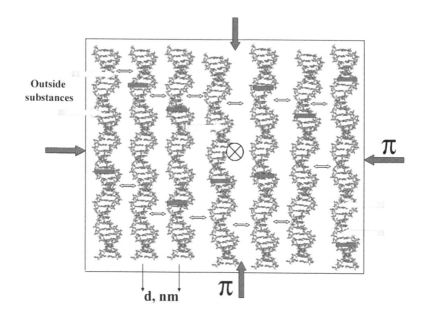

Outside
substances

π

π

d, nm

π

FIGURE 3.13 Structure of a quasinematic layer of DNA molecules in CLCD particles formed by phase exclusion of the molecules.

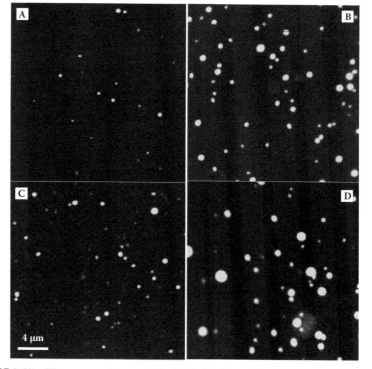

FIGURE 3.19 Fluorescence images of the DNA CLCD particles treated with SYBR Green.

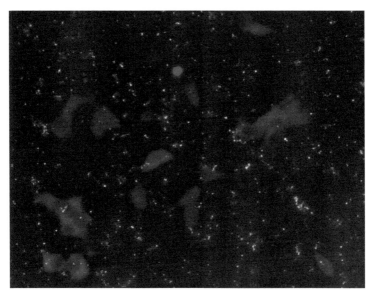

FIGURE 3.20 Images of DNA CLCD particles (green fluorescence) and cells transfected by recombinant plasmid expressing a red protein.

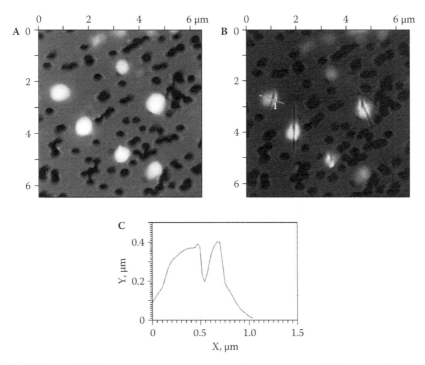

FIGURE 4.6 AFM images of rigid nanoconstructions formed by DNA molecules cross-linked with nanobridges and the profile of the cut of the nanoconstruction.

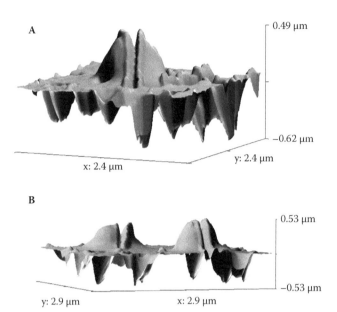

FIGURE 4.7 Spatial images of cuts of rigid nanoconstructions formed by DNA molecules cross-linked with nanobridges.

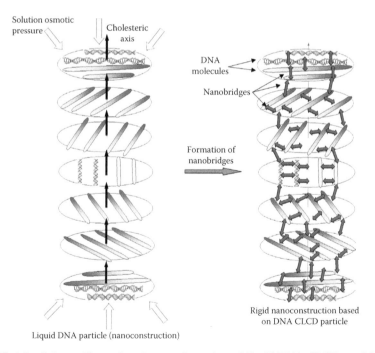

FIGURE 4.8 Scheme illustrating the transformation of liquid DNA CLCD particle into rigid DNA nanoconstruction.

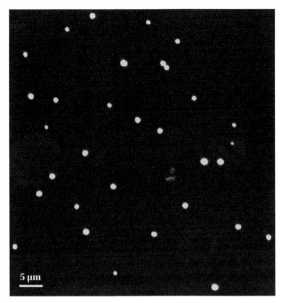

FIGURE 4.18 Image of CLCD particles of [DNA-Gd^{3+}] complexes in water–salt PEG-containing solution treated with SYBR Green. (Image was taken by Leica TCS SP5 confocal microscope.)

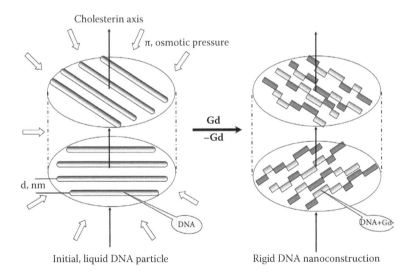

FIGURE 4.21 Scheme of transition of the structure of CLCD particle from a liquid to a rigid state induced by a high concentration of gadolinium ions.

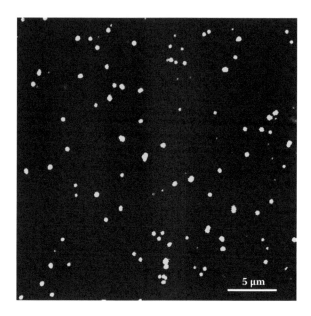

FIGURE 4.25 Fluorescence image of CLCD particles formed by DNA in PEG-containing solution and treated with Au nanoparticles (2 nm) and then with SYBR Green.

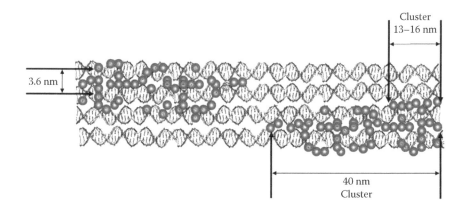

FIGURE 4.26 Hypothetical model of location of Au clusters between double-stranded DNA molecules forming the quasinematic layer.

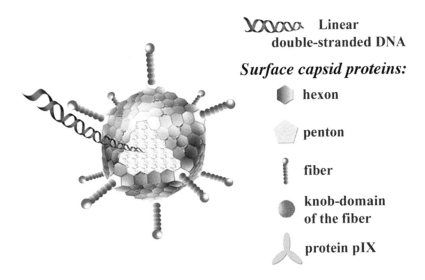

FIGURE 5.1 Structure of adenovirus virion.

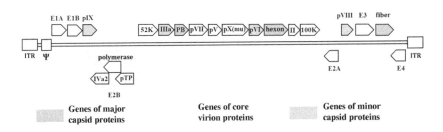

FIGURE 5.2 Schematic representation of adenovirus genome. Legend keys: ITR, inverted terminal repeat; P, protease; Ψ, packing signal; PB, penton base.

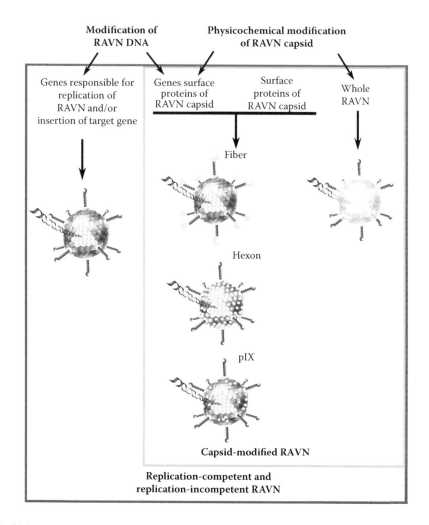

FIGURE 5.3 Strategies for RAVN construction.

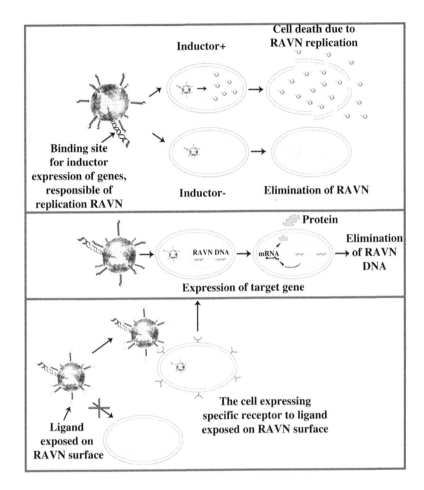

FIGURE 5.4 Effects of RAVNs with modified genomes: (A) replication-competent; (B) replication-incompetent; (C) capsid-modified.

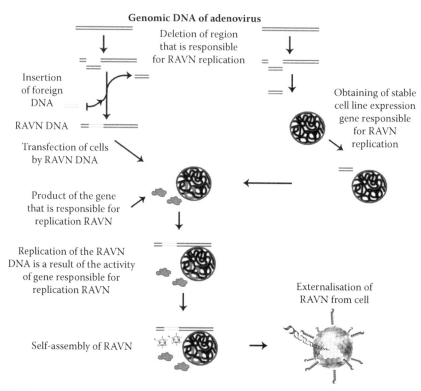

Genomic DNA of adenovirus

Deletion of region that is responsible for RAVN replication

Insertion of foreign DNA

RAVN DNA

Transfection of cells by RAVN DNA

Obtaining of stable cell line expression gene responsible for RAVN replication

Product of the gene that is responsible for replication RAVN

Replication of the RAVN DNA is a result of the activity of gene responsible for replication RAVN

Externalisation of RAVN from cell

Self-assembly of RAVN

FIGURE 5.5 Formation of replication-incompetent RAVNs.

FIGURE 6.5 Piece of a polymeric hydrogel containing DNA CLCD particles pretreated with SYBR Green fluorescent dye.

3 "Liquid" Nanoconstructions Based on Spatially Ordered Double-Stranded DNA Molecules

3.1 DOUBLE-STRANDED DNA MOLECULES AS BUILDING BLOCKS FOR NANOTECHNOLOGY

As noted in chapter 1, the contemporary bottom-up approach in nanotechnology is usually targeted at the assembly of spatial functional structures (devices) from desired building blocks. If these components are based on biological molecules (nucleic acids, proteins, etc.), they must satisfy certain requirements. First, they need to have definite physical (optical, electronic, etc.) and chemical (catalytic, etc.) properties that make it possible to both easily control the structure of the created nanometric biological structures and apply them practically. Second, the properties of the initial building blocks (modules) have to be predictable and programmable based on the specific compositional, configurational, conformational, and dynamical properties of the biological molecules forming the modules, thereby providing the simplicity and efficiency needed to form spatial nanostructures. Third, the building blocks must have a certain size to fill the gap between the submicrometer sizes reached with the help of the classical top-down technology and the sizes that can be reached by classical bottom-up technologies, such as chemical synthesis or supermolecular self-assembly. Fourth, there must be simple methods of regulating the properties of the created nanostructures and controlling the changes in these properties. Finally, the nanostructures have to be able to carry high concentrations of "guest" molecules and determine their location in the created object.

Enumeration of these requirements leads to the assumption that the creation of spatial nanostructures requires the use of a method other than the hybridization technique described in chapter 2, which involves the tailored ordering of adjacent single-stranded nucleic acid molecules. This suggests that spatially ordered structures can only be easily formed using a new technique based on the application of double-stranded nucleic acid molecules and their complexes—somehow fixed in space relative to each other—and not the individual single-stranded molecules as described in chapter 2. In particular, it is well known that spatial structures (condensed phases) are formed as a result of the phase exclusion (condensation) of double-stranded nucleic acid molecules from water–salt solutions. Such a technique for creating extensional nanometric structures has been developed at the Institute of Molecular Biology Institute and is called *DNA nanoconstruction.*

The properties of *single-stranded* DNA molecules for use as building blocks to create nanostructures using a hybridization approach were discussed in chapter 2. The properties of *double-stranded* DNA molecules that are important for the implementation of a new technique for creating nanostructures can also be enumerated as follows:

1. Double-stranded DNA molecules with molecular mass below $1–3 \times 10^6$ g/mol, i.e., molecules that correspond to the concept of nanometric objects, have a rigid structure that usually matches the B form.
2. The properties of rigid DNA molecules and the type of intermolecular interactions between these molecules in various conditions are predictable.
3. Nitrogen bases retain their chemical reactivity in DNA molecules, forming spatial structures by phase exclusion.
4. The phase exclusion of rigid, linear, double-stranded DNA molecules of any origin (regardless of their nucleotide composition and their source) at certain conditions leads to the formation of condensed phases (or dispersions) of these molecules with specific physicochemical properties.

These enumerated peculiarities demonstrate that double-stranded DNA molecules have predictable properties and programmable intermolecular interactions.

There is another peculiarity that attracts interest to the condensed phases of DNA. Considering the known properties of one of the condensed forms of polymers, namely the lyotropic liquid crystals (LC), attention can be directed to a number of features that apparently have much in common with the properties of DNA molecules in biological objects. First, some of the mechanisms of the formation of LC are realized practically without energy losses and lead to almost 100% yield of the final product. Second, the small-angle X-ray scattering parameters of DNA molecules in a condensed phase are relatively close to the parameters of DNA molecules in biological objects such as DNA-containing viruses or protozoan chromosomes. Third, there are multiple LC phases, with the transition between them regulated by both the properties of the DNA and the medium in which the LC are formed.

Linear, rigid, double-stranded native DNA has two peculiarities that determine many of the physical properties of the condensed state of DNA. The issue is that DNA molecules have both geometric anisotropy (they are helical) and optical anisotropy (determined by the presence of an asymmetric carbon atom in the sugar residues). These two special features cannot be changed, as they are interior properties typical of low-molecular-mass DNA molecules under any conditions. In the language of physics and physical chemistry, it means that these double-stranded DNA molecules approaching each other will tend to the so-called cholesteric (helically twisted) packing, which determines the easily detected and most specific optical properties of the DNA cholesteric phase. This specific property of double-stranded DNA molecules cannot be removed; it can only be affected by making the DNA molecules or their complexes change the type of packing. However, the tendency of rigid double-stranded (but not flexible single-stranded) DNA molecules to form a cholesteric phase will realize itself at any opportunity, where it will dominate and determine the properties of the DNA condensed phase.

A number of additional peculiarities of LC DNA pose some interest from a biological point of view. On the one hand, a targeted modification of the secondary structure of DNA molecules in an LC phase at a relatively broad scope does not lead to any changes in the type of spatial packing, i.e., an LC system has "structural memory" determined by the properties of the whole ensemble of molecules, not individual DNA fragments. On the other hand, at the moment of the formation of LC, when DNA molecules or their complexes, for instance, with histone proteins, "recognize" each other, the properties of these molecules exert a most significant, if not crucial, influence on the type of spatial packing. At the same time, even minor changes in the structure of individual DNA can regulate the spatial structure of the whole ensemble of molecules. Thus, there are many similarities between the peculiarities of DNA in cells and the properties of LC DNA that attracts even more interest to the application of LC DNA as a background for nanoconstruction.

The assumption that the formation of LC structures of DNA can be a basis for creating spatial nanostructures was stated in the work of Yevdokimov, Skuridin, and Salyanov [1], while the first attempts to create LC DNA–based nanoconstructions were described in the work of Yevdokimov et al. [2]. It can be noted that the opinion that the application of LC DNA structures will make it possible to meet the requirements for spatial nanoconstructions was repeated in works not only by Seeman [3], one of the founders of the hybridization technique, but also by other authors [4]. It is interesting that a number of works by various authors [5–8] published in 2009 formulated the assumption that layers of adjacent double-stranded DNA molecules packed the same way as the molecules in LC phases—and not single-stranded DNA molecules (or DX tiles)—could be used as building blocks for DNA nanoconstructions. These works also described the creation of three-dimensional (3-D) DNA nanostructures based on LC DNA.

These findings imply that before using DNA molecules spatially fixed in an LC phase (dispersion) as building blocks, it is necessary to determine the fundamental parameters of LC structures. It has to be noted that, unlike most of the limited sets of methods used to analyze DNA nanostructures obtained by the hybridization technique, in this case there is a broad range of simple physicochemical techniques that allow one to control the formation of DNA LC phases (dispersions) and analyze even minor changes in the structure of these phases.

A great deal of attention was paid in the literature [9] to studies of the formation of phases (phase exclusion) of double-stranded nucleic acids (and synthetic polynucleotides) of low molecular mass as a result of so-called entropy condensation (entropy ordering). The phase exclusion of nucleic acids from water–polymer solutions can also lead to the formation of dispersions of these molecules, while the dispersion particles are characterized by ordered spatial packing of nucleic acid molecules.

The question of the properties of such DNA dispersions is interesting from at least two points of view. From the theoretical point of view, the interest in the dispersions is aroused by the fact that the physicochemical properties of dispersion particles below 100 nm may be significantly different from the typical properties of massive (continuous) phases [10]. As shown in chapter 1, the differences in properties can be ascribed to the size effects typical of nanometric objects. Such effects are determined by the existence of energy contributions into the free energy made by the surface tension of particles and by the presence of possible defects in the packing of molecules in the particles. From a biological point of view, properties of DNA dispersions are interesting because such particles make it possible to describe more accurately the properties of DNA-containing viruses or chromosomes, which are actually microscopic dispersion systems characterized by both ordered and flexible packing of DNA.

The properties of nucleic acid liquid crystals and, especially, DNA liquid-crystalline dispersions were thoroughly described in our book *DNA Liquid-Crystalline Dispersions and Nanoconstructions* (CRC Press), published in 2011. Thus the following discussion will highlight only the most important properties of these dispersions.

3.2 PHASE EXCLUSION OF DOUBLE-STRANDED DNA MOLECULES FROM POLYETHYLENE GLYCOL SOLUTIONS

Mixing water–salt solutions of nucleic acids, namely, DNA with water–salt solutions of neutral water-soluble synthetic polymers, namely, polyethylene glycol (PEG), leads to the formation of nucleic acid dispersions [11]. Such condensations are known as ψ-condensations of DNA. (The Greek letter ψ [psi] is a symbol for *polymer-salt-induced.*) PEG molecules are not included in the composition of the formed dispersion particles.

The scheme (Figure 3.1) of condensation of DNA with molecular mass $0.6–0.8 \times 10^6$ g/mol, i.e., molecules with a length of about 200–400 nm, shows that, though the efficiency of the formation of a DNA dispersion is affected by a number of variables, these variables can easily be controlled and regulated according to the goal of the experiment.

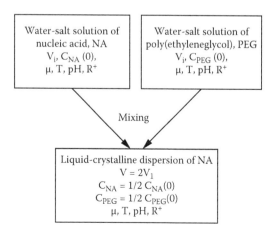

FIGURE 3.1 Scheme of the formation of a dispersion from double-stranded nucleic acid molecules.

The easiest way to detect a dispersion formation is registration of its absorption spectrum. Starting from a particular (critical) concentration of PEG (C^{cr}_{PEG}) in solution, the optical density in the absorption band of DNA nitrogen bases decreases, while at wavelengths above 320 nm (where neither DNA nor PEG molecules absorb) it increases. (See Figure 3.2, where: curve 1 $C_{PEG} = 0$; curve 2 $C_{PEG} = 150$ mg/mL; curve 3 $C_{PEG} = 180$ mg/mL; curves 4 and 5 are curves 2 and 3 minus the apparent (A_{app}) optical density ($A_{app} = K\lambda^{-n}$), where $n = f(R)$. The inset shows the dependence of A_{app} ($\lambda = 340$ nm) on PEG concentration, where PEG molecular mass = 4,000 g/mol, 0.3 mol/L NaCl.) The increase in optical density in this area of the spectrum is caused by the dispersion of light scattering due to formation on the DNA dispersion particles (an apparent optical density, A_{app}).

The theoretical calculations based on the measurements of optical, hydrodynamic, and other physicochemical properties of the dispersions combined with application of certain theoretical approximations make it possible to evaluate the parameters of DNA dispersion particles necessary to estimate the molecular mass of the particles formed.

The results obtained by various physicochemical methods show that rigid, linear, double-stranded DNA molecules with low molecular mass (as a result of the phase exclusion from PEG-containing solutions) form dispersions with size (D) of the particles $\approx 10^2$ nm. An increase in the DNA concentration in the solution or in the molecular mass of the DNA samples leads to an increase in D value, while an increase of PEG concentration leads to a decrease in D value.

Evaluation of the molecular mass of DNA dispersion particles based on the obtained values of the sedimentation coefficient (S) and the coefficient of translation diffusion (D_t) at $C_{DNA} \to 0$ shows that the molecular mass of one particle is close to 5×10^{10} g/mol. As the molecular mass of DNA molecules used for the formation of the dispersion is about 10^6 g/mol on average, it can be considered that one dispersion

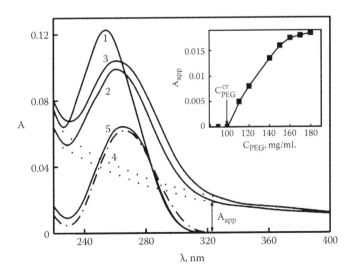

FIGURE 3.2 Change in the DNA absorption spectrum at the formation of dispersion in PEG water–salt solution.

particle formed at C_{PEG} = 170 mg/mL contains approximately 10^4–10^5 DNA molecules [12]. The explanation for the size of such particles of a DNA dispersion in PEG-containing solution considers the contributions of both kinetic and thermodynamic factors. One can stress that these evaluations are approximate, as they are based on the application of methods considering DNA dispersion particles to be spherical objects.

The registration of the absorption spectra allows one to evaluate the critical PEG concentration (C^{cr}_{PEG}) at which the formation of a DNA dispersion takes place (inset in Figure 3.2). By the definition, the critical concentration of PEG is the concentration at which its coils overlap [13]. The overlap of PEG coils that takes place as PEG concentration increases is an important factor that contributes to the phase division (otherwise, the compressing osmotic pressure of PEG is insufficient [13, 14]). The order of the concentration of c* segments (at which the overlap takes place) that must correspond to the experimentally detected critical concentration of PEG (C^{cr}_{PEG}) was calculated and compared with the last value.

The results illustrated in Figure 3.3 show that, first, the C^{cr}_{PEG} value is inversely proportional to the molecular mass of PEG preparations [15]; second, the formation of a DNA dispersion takes place at PEG concentrations that match the c* value only by order. (In Figure 3.3, curve 1 is 0.3 mol/L NaCl; curve 2 is 0.3 mol/L CsCl. The dotted curve 3 is the dependence of the c* value based on the calculation of the hydrodynamic properties of PEG solutions.) This proves the approximate character of the evaluation and the fact that the effects of such parameters as solution ionic strength, the hydrate shell around PEG molecules, and so forth, on the PEG structure is not considered. Figure 3.4 shows as well that the molecular mass of PEG at which a DNA dispersion forms must be above 600 g/mol [16].

The two principal results—namely, the existence of the dependence between the critical concentration of PEG and PEG molecular mass, as well as the formation of a DNA dispersion when PEG molecular mass is above the critical value—were used as the background to the theory proposed by Grosberg [17] to describe the

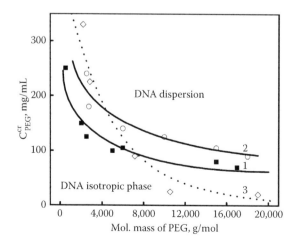

FIGURE 3.3 Dependence of critical concentration of PEG on the PEG molecular mass.

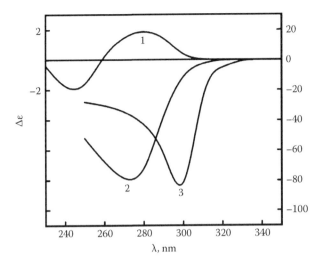

FIGURE 3.4   CD spectra of a water–salt solution of a double-stranded B-form DNA, a DNA dispersion, and a thin layer of a cholesteric liquid-crystalline DNA phase.

condensation of DNA molecules. In this theory, PEG molecules not only leave the volume of the formed DNA dispersion particles, but also exert a compressing effect on these particles (which causes the dependence on C_{PEG}).

Therefore, rigid, linear, double-stranded molecules of nucleic acids and synthetic polymers can form dispersions in PEG-containing solutions. The factors that affect the efficiency of the formation of such dispersions are the properties of the solvent (PEG concentration and molecular mass, etc.) as well as those of the DNA molecules (molecular mass, concentration, etc.).

It has to be added that particles of the low-molecular-mass DNA dispersions represent nanosized drops of concentrated DNA solution, which cannot be "taken in hand" or "directly seen." A liquid mode of packing these molecules in particles of dispersions prevents their immobilization on the surface of a membrane filter in an immutable shape. Consequently, the problems surrounding the possibility of "seeing" particles of dispersions and how their dimensions can be directly evaluated have not been solved. The question of how nucleic acid molecules are packed in these particles is also still open.

3.3   THE CHOLESTERIC MODE OF NUCLEIC ACID MOLECULES PACKING IN PARTICLES OF DISPERSIONS RESULTS IN AN APPEARANCE OF ABNORMAL CD SPECTRA

The mode of packing of low-molecular-mass double-stranded DNA molecules in liquid-crystalline phases was determined by analyzing the textures of the phases [9]. However, additional research needs to be performed to determine how these DNA molecules are packed in dispersion particles.

The matter is that the existence of the size effect can lead to noticeable differences between packing of the DNA molecules in dispersion particles and their packing in massive phases. This difference can be especially noticeable in the packing mode and, in this case, requires extra testing. The presence of the size effect can alter the packing typically expected for liquid-crystalline phases in the case of dispersions. So in the case of DNA molecules with low molecular mass (forming dispersions at phase exclusion from water–polymeric solutions), it is necessary determine the character of DNA packing in these particles experimentally. In this connection, the methods that make it possible to answer the question given in the last paragraph of section 3.2 is especially important. One of the most significant methods is the *spectroscopy of circular dichroism* (CD). Here one can add that the theoretical aspects of this method (and especially its application to the study of the DNA dispersion particles) were analyzed earlier in our book *DNA Liquid-Crystalline Dispersions and Nanoconstructions* (CRC Press) and in *The CD Spectra of Double-Stranded DNA Liquid-Crystalline Dispersions* (Nova Sci. Publishers, 2011). Thus we will only highlight the main peculiarities of the CD spectra that are important for DNA nanoconstruction.

Figure 3.4 shows the experimentally measured CD spectra of a solution of an initial, linear, double-stranded DNA molecule (B form, curve 1) and a dispersion formed by phase exclusion of these molecules from water–salt–PEG-containing solution (curve 2) as well as a thin layer (20 μm cell) of cholesteric liquid-crystalline phase [9] (curve 3) formed by the same DNA. (In Figure 3.4, curve 1 applies to the left ordinate and curves 2 and 3 to the right ordinate. C_{PEG} = 170 mg/mL, 0.3 mol/L NaCl, 0.01 mol/L phosphate buffer, pH 6.8.) One can stress here that the formation of optically active dispersions takes place only for water–salt–PEG-containing solutions with high ionic strength providing 90%–95% neutralization of negative charges of the DNA phosphate groups. From the figure, it can be seen that the shape of the band in the CD spectrum of he DNA phase is similar to the shape of the absorption spectrum (Figure 3.2), although the maximum for this band is red-shifted ($\lambda \approx 300$ nm). A comparison of curves 1 and 2 shows that the intense band in the CD spectrum is characteristic of the DNA cholesteric liquid-crystalline phase. The band has a negative sign and is located in the absorption area of DNA nitrogen bases.

In the case of the thin layer of the DNA cholesteric phase, both the wavelength, which corresponds to the maximum of the negative band in the CD spectrum, and the amplitude of the intense (abnormal) band depend on the geometry of the layer, on its position in respect to the polarized light beam, on the thickness of the layer, and finally, on the homogeneity of the layer structure at the site of light beam penetration. For cholesteric liquid-crystalline phases and low-molecular-mass compounds, all of these questions have been thoroughly analyzed in the literature [18, 19].

The question of interest is how to interpret the CD spectra of a DNA dispersion. To answer it, one has to consider a number of facts.

1. There is a negative band in the CD spectrum in the nitrogen bases absorption region (Figure 3.4, curve 2). The circular dichroism evaluated for the negative band in the CD spectrum of the DNA dispersion as well as for the cholesteric phase is greater than that of the molecular circular dichroism ($\Delta\varepsilon \approx 2$ units), which reflects the properties of isolated chromophores

(nitrogen bases absorbing in the UV region of the spectrum) or of nitrogen based within the initial linear DNA molecules. The negative signs of the bands in the CD spectra mean that the adjacent right-hand-twisted DNA molecules have a left-handed helical twist.

2. The investigations of textures of cholesteric phases formed by low- molecular-mass DNA molecules show that the pitch, P, of cholesteric spatial structures is varied within 2,000–3,000 nm [20]. A sharp increase in the band amplitude in the CD spectrum observed at formation of the DNA dispersion could not be explained by the diffraction effect (Bormann effect) that evolves when the region of absorption and selective reflection bands of the cholesteric phase collide. This conclusion is confirmed by the experiments, in the course of which the local value of circular dichroism for relatively small sites (domains) in the film of DNA cholesteric phases was locally measured with the help of the CD microscope [20].

To calculate the CD spectra of a DNA dispersion, a phenomenological model based both on previous theories [21–29] of optical properties of low-molecular-mass compounds and imperfectly absorbing cholesteric liquid crystals [30] makes it possible to examine the dependence of the abnormal optical activity on the size of the particles, and the pitch of the helical structure [31] was applied to the calculation of the CD spectra. In this case, the particles of the DNA dispersion are considered to be polycrystalline objects with random distribution and orientation of particles possessing their own absorption in the UV region of the spectrum due to the presence of nitrogen bases (chromophores) in the content of DNA molecules. (It is necessary to stress here that the circular dichroism basically reflects the long-range order quasinematic layers formed by neighboring chromosomes, in our case, nitrogen bases.) It is also assumed that individual DNA dispersion particles are quite small to justify the application of the kinematic approximation of the theory of diffraction to describe the optical properties of each particle. The approach based on these assumptions was previously applied to imperfectly nonabsorbing liquid crystals [32].

The theory takes into account both the layered structure of the packed double-stranded DNA molecules and known microscopic data previously shown for these particles. In particular, experimental data enable one to treat the particles of dispersions formed by DNA molecules of low molecular mass (about 10^6 g/mol) as spheres of diameter, D, for which, because of the inherent rigidity of the DNA structure, the liquid-crystalline ordering is specific.

Without going into further details, one can consider only the results of the theoretical calculations [33] that are directly related to the optical properties of DNA dispersion particles and are necessary for the formation of DNA-based nanoconstructions. These results are:

A. We assume that the low-molecular-mass double-stranded DNA molecules are ordered in dispersion particles, as they are in the massive cholesteric liquid-crystalline phase, and, moreover, that the particles are characterized by helically twisted (cholesteric) packing of these molecules, and, hence, that helically twisted packing of quasinematic layers are formed by nitrogen bases (chromophores). If these assumptions are true, then the anisotropy of absorption of nitrogen bases in the content of the DNA molecules for the linear polarized light must show itself as an *intense band in the CD spectrum*. Because orientation of nitrogen bases in the content of neighboring DNA molecules and, hence, in adjacent neighboring layers, is rigidly fixed, one can say that the sign of the band in the CD spectrum in the region of absorption of nitrogen bases will depend on the orientation of the planes of nitrogen bases in respect to the long axis of the DNA molecule.

Figure 3.5 shows the theoretically calculated CD spectrum of a DNA dispersion formed in a PEG-containing solution (C_{DNA} = 10 µg/mL; $\Delta A = (A_L - A_R) \times 2.5 \times 10^{-5}$ optical units). One can see that, in accordance with the foregoing assumptions, the formation of a dispersion as well as a cholesteric phase is followed by an appearance of negative band located in the region of absorption of nitrogen bases.

B. The shape of the band in the CD spectrum of the DNA dispersion is similar to the shape of the absorption band, while the peak of the band in the CD spectrum is red-shifted.

It can also be concluded from Figure 3.5 that the value of molecular circular dichroism (expressed as $\Delta\varepsilon$) calculated for the negative band in the CD spectrum of the DNA dispersion is about 200 units, and it is much larger than that of the value characterizing the molecular circular dichroism of nitrogen bases within the initial linear DNA molecules ($\Delta\varepsilon \approx 2.0$ units). To stress the difference between molecular and a so-called structural circular dichroism, the term *abnormal band* was used to signify an intense band in the CD spectrum [9]. With an allowance for this, the use of the amplitude of this band in the CD spectra, expressed simply as an experimentally measured ΔA value ($\Delta A = [A_L - A_R]$), is more reasonable. This value will be used in our book very often.

According to the foregoing theoretical calculations, the abnormal band in the CD spectrum—located in the absorption region of chromophores incorporated in the content of the massive liquid-crystalline phase—is direct evidence of the formation

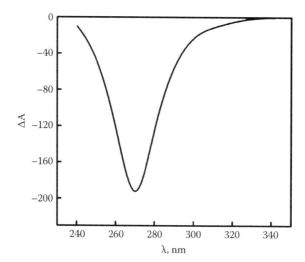

FIGURE 3.5 Theoretically calculated CD spectrum of DNA CLCD.

of a helically twisted structure. Besides, the sense of helicity is reflected in the sight of the abnormal band in the CD spectrum (i.e., a positive band reflects a right-handed helical twist, and a negative band reflects a left-handed arrangement of neighboring quasinematic layers formed by DNA molecules). Random arrangement of high-molecular-mass DNA molecules (for instance, at aggregation of these molecules) or hexagonal packing of low-molecular-mass DNA molecules was unaccompanied by any considerable change in the amplitude of the CD band, except those produced in the secondary structure.

An appearance of an abnormal band in the CD spectra of the DNA cholesteric liquid-crystalline phase and in the dispersion (Figure 3.5) means that the purine and pyrimidine nitrogen bases do play the role of chromophores, providing information about the mode of spatial packing of DNA molecules ordered both in the massive phase and in the particles of dispersions.

Abnormal bands in the CD spectra of both the massive cholesteric liquid-crystalline phase and the DNA dispersion have the same signs (Figure 3.5). The appearance of negative abnormal bands in the CD spectra unequivocally reflects the left-handed twist of neighboring quasinematic DNA molecules spatially packed not only in the massive phase but also in the particles of dispersions (Figures 3.4 and 3.5). Hence, one can conclude that that the right-handed DNA molecules (B form) packed in particles of dispersions form quasinematic layers with a left-handed cholesteric twist.

One can recall that double-stranded DNA molecules are ordered in the particles at distances of 3.0–5.0 nm, that is, they acquire the properties of a crystal; but molecules in the neighboring quasinematic layers are mobile; that is, they retain the properties of a liquid. To stress this fact, the term *liquid-crystalline dispersions* (LCD) is used to signify these dispersions. To stress the spatial twist of quasinematic layers, the term *cholesteric liquid-crystalline dispersions* (CLCD) is used to signify these dispersions [33]. This conclusion corresponds to the general idea that the cholesteric packing is a specific property of any molecules with inherent geometrical and/or optical anisotropy, including double-stranded nucleic acid molecules (DNA, RNA), which realize their tendency to such a packing mode at any opportunity.

C. The value of the amplitude of the abnormal band in the CD spectra of DNA CLCD depends both on the size of the particles and on the pitch, P, value of the cholesteric structure.

Figure 3.6 shows the theoretical CD spectra of DNA cholesteric liquid-crystalline dispersions (CLCD) whose particle size varies (100, 200, 300, 400, and 500 nm, respectively, for curves 1–5) while the pitch of the cholesteric helix is constant ($P = 2,000$ nm) and $\Delta A = (A_L - A_R) \times 10^{-5}$ optical units. The amplitude of the abnormal band depends on the diameter of CLCD particles and increases as the diameter increases (inset in Figure 3.6).

Besides, the calculation shows that, if the diameter of CLCD particles reaches the minimal value of 50 nm, the amplitude of the abnormal band in the CD spectrum decreases so sharply that it no longer can be distinguished from that of the CD spectrum typical of the initial linear DNA [9].

Fixing the size of the DNA CLCD particles (assuming, for instance, that $D = 400$ or 500 nm), the effect of the pitch, P, value of the cholesteric structure on the amplitude of the band in the CD spectrum can be evaluated. The calculated CD spectra of

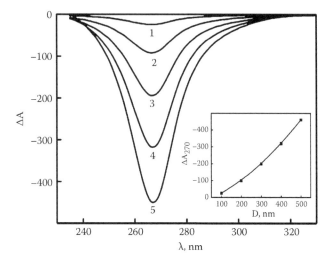

FIGURE 3.6 Theoretically calculated CD spectra of DNA CLCD with different sizes of particles.

DNA CLCD whose particles are characterized by cholesteric twist with a different pitch, P, value are shown in Figure 3.7 ($P = 2,000$; $4,000$; $6,000$; $8,000$; and $10,000$ nm for curves 1–5, respectively; $D = 500$ nm) and $\Delta A = (A_L - A_R) \times 10^{-5}$ optical units). It can be noticed that the smaller P value of the DNA cholesteric structure (i.e., the greater the twist—the angle between the adjacent quasinematic layers of DNA in the helical structure cholesteric), the more intense is the band in the CD spectrum, and vice versa. (The inset in Figure 3.7 represents the dependence of ΔA ($\lambda = 270$ nm) on the P value of the cholesteric helix.)

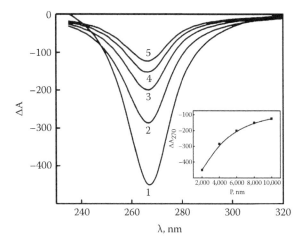

FIGURE 3.7 Theoretically calculated CD spectra of DNA CLCDs whose particles have different values of cholesteric pitch.

The theoretical treatment has also shown that at a P value of about 30 μm and with the constant structural properties of the DNA molecule, the amplitude of the band in the CD spectrum is quite close to the amplitude of the band characteristic of isolated linear DNA molecules (Figure 3.7).

D. The sign of the band in the CD spectra of DNA CLCD depends on the sense of the twist of the spatial cholesteric structure formed by DNA molecules packed in a CLCD particle. Basically, the result of the theoretical calculations means that the change in the direction of the twist of the cholesteric packing of DNA molecules leads to the change of the negative sign of the abnormal band to a positive one, while the shape of the CD spectrum does not change.

E. The role of the length of DNA molecules in the formation of CLCD particles is also important (Figure 3.8). (In Figure 3.8, the preparations of nucleosome DNA from chromatin of Ehrlich ascites tumor cells have been used; molecular mass of PEG = 4,000 g/mol; C_{PEG} = 170 mg/mL, 0.3 mol/L NaCl.) Unlike aggregation of the common DNA molecules, which can occur at any length of molecules and is not followed by the appearance of abnormal optical activity, the DNA CLCD is formed only in cases when the length of the molecules is above 15 nm (≈50 base pairs). The obtained result demonstrates that there is a lower limit in DNA length at which the optically active dispersion is formed. (One can stress here that obtaining homogeneous DNA preparations with small lengths is a very difficult experimental task. For purely technical reasons, it is better to use the DNA preparation with higher molecular masses.)

The upper DNA molecular mass limit at which the molecules maintain the ability to form CLCD is close to 3×10^6 g/mol [20, 21]. At DNA molecular mass greater than 10×10^6 g/mol, the CD spectra of the formed dispersions differ only slightly from the CD spectra of the linear DNA.

The role of the secondary structure of nucleic acid molecules attracts special interest. Only rigid, linear, double-stranded DNA (RNA) molecules can form dispersions

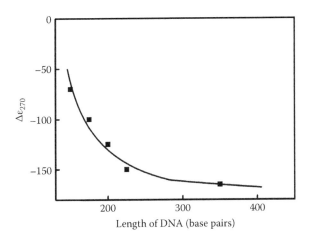

FIGURE 3.8 Dependence of $\Delta\varepsilon_{270}$ value for CLCD particles formed from DNA molecules with different lengths.

with cholesteric packing of these molecules in the CLCD particles by phase exclusion. The formation of dispersions from flexible single-stranded nucleic acid molecules with less rigid structure combined with distorted orientation of nitrogen bases does not lead to the appearance of an abnormal band in the CD spectra. The CD spectra of CLCD formed by double-stranded molecules of various polynucleotides show that even minor changes in the structure of the double helix are enough to cause a change in both the amplitude and the sign of the abnormal band in the CD spectra of these dispersions. For instance, this happens in the case of poly(dA-dT)×poly(dA-dT) and poly(dA)×poly(dT) molecules. These synthetic polymers form CLCD whose CD spectra are mirror images of each other. A similar effect was noticed at the transition from DNA CLCD to CLCD of double-stranded RNA. Namely, in water–salt solution with a high ionic strength ($\mu \approx 1.0$), right-handed twisted molecules of the replicative form of double-stranded f2 phage RNA can form CLCD with both positive and negative abnormal bands in the CD spectra under an increase in PEG concentration in solution.

F. The process of DNA (RNA) CLCD particle formation depends on the osmotic pressure of the solution as determined by both the concentration of PEG in solution and its molecular mass. When the PEG concentration in water–salt solution with a moderate ionic strength ($\mu \approx 0.15$–0.3) increases, the DNA CLCD formation is followed by the appearance of an abnormal band in the CD spectrum when the critical PEG concentration (C^{cr}_{PEG}, compare it with Figure 3.2) is reached, but this abnormal band disappears as the limit PEG concentration ($C^{lim}_{PEG} \approx 260$–$300$ mg/mL) in solution is reached. (See Figure 3.9, where curve 1 is NaCl; curve 2 is NaClO$_4$; and curve 3 is CsCl. PEG molecular mass = 4,000 g/mol, and salt concentration = 0.3 mol/L.) This figure shows that at low concentrations of PEG ($C_{PEG} < 150$ mg/mL), DNA molecules are in an isotropic state, and there is only a low-intensity band in their CD

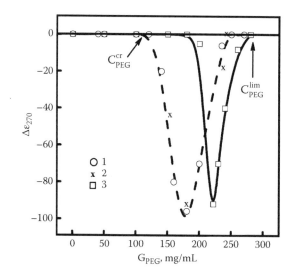

FIGURE 3.9 Dependencies of the amplitude of the abnormal negative band in the CD spectra of DNA ($\lambda = 270$ nm) on the PEG concentration in solutions of different salts.

spectra, typical of the DNA B form (see Figure 3.4, curve 1). At $C_{PEG} > 300$ mg/mL, dispersions are still formed, but they do not possess any abnormal optical activity. Hence, the DNA molecules can form CLCD only within an area of concentrations limited by C^{cr}_{PEG} and C^{lim}_{PEG} values. Therefore, as a result of the phase exclusion performed at certain critical conditions, the rigid double-stranded DNA molecules with low molecular mass can be ordered to form CLCD particles. Crossing of the lower limit causes a transition to the isotropic state of these molecules, while crossing of the upper limit leads to a transition where the particles show no abnormal optical activity.

However, the condensation of DNA molecules, by itself, is not a sufficient condition for an appearance of intense band in the CD spectrum, since many aggregated forms of DNA (for instance, the aggregates formed by single-stranded DNA molecules) fail to show the intense band in the CD spectra. Therefore, on one hand, an appearance of an intense band in the CD spectrum is connected with long-range coupling of the dipoles of nitrogen bases, i.e., the nitrogen bases are main contributors into the optical behavior of the CLCD (in the amplitude of an intense band in the CD spectrum). On the other hand, taking into account that nitrogen bases are fixed rigidly enough in the secondary DNA structure, an appearance of this band reflects the specific type condensation of the double-stranded DNA molecules, i.e., the helical array of these DNA molecules.

By a helical array, one can consider a parallel organization of double-stranded DNA molecules that is then twisted slightly so that each quasinematic layer of DNA molecules is at a slight angle of twist with respect to the two neighboring ones. As the DNA molecules condense into particles of dispersion, the local density of nitrogen bases increases to the point where significant delocalization can occur. This means that the amplitude of an intense band in the CD spectra of the DNA CLCD, formed by DNA molecules that possess native secondary structure and fixed properties of nitrogen bases, is connected both with "long-range collective behavior" of the nitrogen bases and with a local density of DNA molecules in the particles of CLCD. This, in turn, depends on the secondary structure of these molecules and on the properties on the solvent.

This means that the amplitude of an intense band in the CD spectra of the DNA CLCD, formed by DNA molecules that possess native secondary structure and fixed properties of nitrogen bases, is connected both with long-range collective behavior of the nitrogen bases and with a local density of double-stranded DNA molecules in the particles of CLCD. This, in turn, depends on the secondary structure of these molecules and on the properties of the solvent. This means that the amplitude of intense band in the CD spectra can vary over a broad range of values. Hence, the value of $\Delta\varepsilon$ is not a constant. Therefore, the use of $\Delta\varepsilon$ values may have only an illustrative character, and it seems illogical to apply this parameter to evaluate the peculiarities of the CD spectra of various DNA CLCD.

To stress the difference between molecular and a so-called structural circular dichroism, the term *abnormal band* was used to signify an intense band in the CD spectrum [33]. Given the foregoing considerations, the use of the amplitude of this band in CD spectra, expressed simply as an experimentally measured ΔA value, is more reasonable than $\Delta\varepsilon$. Consequently, the ΔA value will be used in our subsequent discussions.

G. Temperature increases the diffusion mobility of DNA molecules in the particles of the CLCD and initiates the transition of their spatial structure into an optically inactive state. As the temperature of the PEG-containing solution increases, the amplitudes of the abnormal bands in the CD spectra of various nucleic acid CLCD decrease, regardless of their signs. (See Figure 3.10, where PEG molecular mass = 4,000 g/mol; curve 1 is 80 mg/mL PEG; curve 2 is 100 mg/mL PEG; curve 3 is the melting curve of the secondary structure of poly(dA)×poly(dT) registered by changes in the optical density of the dispersion at 260 nm.) When the temperature of a PEG-containing solution increases, its osmotic pressure does not change significantly, and the separation of the strands of DNA molecules in particles is impossible due to steric reasons [34]. The process that reflects this decrease in the amplitude of the abnormal band in the CD spectra is called CD-melting [35], and is similar to the melting of the secondary structure of double-stranded nucleic acids (denaturation). CD-melting is characterized by the temperature of the transition varying from 30°C to 80°C (depending on the PEG concentration and nucleotide composition of nucleic acid molecules); it is usually marked as τ_m, similar to the melting temperature (T_m) of nucleic acids' secondary structure.

In Figure 3.10, the melting curves of the CLCD particles formed by poly(dA)×poly(dT) molecules under different conditions and detected by various methods are compared. It can be noticed that, in the case of the polynucleotides dispersions, the amplitude of the positive band in the CD spectra (curves 1–2) decreases and, at a certain temperature, the band disappears. Meanwhile, the CD spectrum becomes similar to the polynucleotide spectrum under the absence of PEG. At

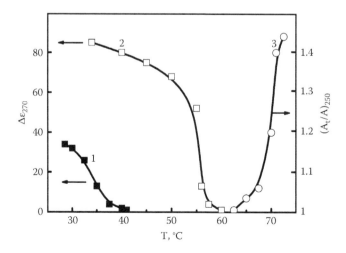

FIGURE 3.10 Dependence of the amplitude of the abnormal positive band in the CD spectra of CLCD formed in PEG-containing solutions by poly(dA) × poly(dT) molecules on the temperature.

relatively low values of C_{PEG}, the abnormal band in the CD spectrum disappears at a temperature lower than the temperature of polynucleotide denaturation (curve 1).

The cooling of the PEG solution after CD-melting always results in full restoration of, and even an in increase in, the abnormal band amplitude in the CD spectrum of the CLCD; hence, initial cholesteric packing of DNA molecules in CLCD particles is restored [34–36]. (This means that the decrease in the amplitude of the abnormal band in the CD spectrum reflects the untwisting of the cholesteric structure of dispersion particles.) Such thermal training is often used to achieve the maximal optical activity of CLCD of different nucleic acids.

As for the strand separation (denaturation) of double-stranded polynucleotides, as well as DNA and RNA contained in CLCD particles, the following fact should be noted: The destruction of the secondary structure of these polynucleotides or nucleic acids (melting) within CLCD particles is followed by the hyperchromic effect, while the optical parameters of the melting of these molecules with or without PEG almost coincide. The hyperchromic effect corresponding to the denaturation of these molecules within CLCD particles is observed at a temperature ≈85°C. The precise value of this temperature depends on the nucleotide composition of the polynucleotides and only slightly (by 2°C–7°C) increases with an increase in the PEG concentration used to prepare the dispersions [35]. The relatively high percentage of renaturation (80%–100%) registered by the changes in CLCD absorption spectra should also be marked.

H. The fact that the abnormal band in a CD spectrum typical of a thin layer of massive DNA cholesteric liquid-crystalline phase exists in the case of DNA CLCD made it possible to determine the parameters of the short-range order of DNA molecules in the phases that can be obtained from CLCD particles as a result of their low-speed centrifugation. The study of small-angle X-ray scattering from such phases formed at a certain PEG concentration (i.e., in solutions with a certain osmotic pressure [1]) allows one not only to verify the assumption about the ordered state of DNA molecules in CLCD particles, but also to evaluate DNA concentration in the CLCD particles.

The X-ray parameters of these phases and the CD spectra of dispersions pointing to the existence of dispersions with different optical properties permit one to compare the borders for the existence of the main packing types of DNA molecules in massive phases and in dispersions (Figure 3.11).

The area of existence of DNA CLCD particles is shaded in Figure 3.11. The second horizontal axis corresponds to the DNA concentration in phases formed as a result of low-speed sedimentation of CLCD particles during their low-speed centrifugation. It is notable that the CLCD particles are formed at the solution osmotic pressure that provides a distance of 3–5 nm between DNA molecules. At the PEG solution osmotic pressure that reduces the distance between DNA molecules to 2.9 nm or less, these molecules form a hexagonal phase that has no abnormal optical activity, while the right-handed helical twist and initial secondary structure of the individual nucleic acid molecules is still maintained. Figure 3.11 shows as well that the local concentration of DNA typical of CLCD particles varies within 180–400 mg/mL. It should be highlighted again that the packing of DNA molecules in dispersion particles has a liquid nature, and DNA molecules in every quasinematic layer can both rotate around their axes and be displaced laterally.

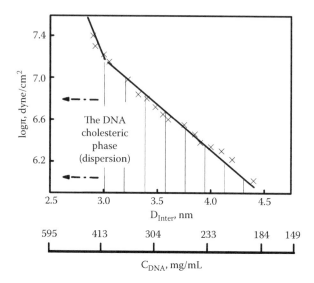

FIGURE 3.11 Dependence of the average distance between DNA molecules in phases formed in PEG-containing solution on the osmotic pressure of the solution.

In conclusion, one can say the following: The theoretical analysis—based on experimental data received by various methods and describing the process of the phase exclusion of double-stranded molecules of nucleic acids (DNA and RNA)—makes it possible to represent the formation of the CLCD particles of these molecules as seen in Figure 3.12. (Here, the CLCD particle is shown as an oval; the structure of a quasinematic layer with distance, d, between adjacent DNA molecules is also shown. In the case shown here, the diameter of a particle is approximately 500 nm because the molecular mass of the used DNA molecules was $(6–8) \times 10^5$ g/mol.)

It follows from Figure 3.12 that the structure of particles of dispersion formed at the phase exclusion of DNA molecules from water–polymeric solution is a combination of so-called quasinematic layers, where adjacent DNA molecules are ordered along each other, while every layer is turned at a certain angle relative to the next one. The twist of the layers leads to the helically twisted (cholesteric) structure of the whole particle.

As seen in the graph at the right-hand side of Figure 3.12, the formation of DNA CLCD is followed by the appearance of the abnormal negative band, as a rule, while the formation of RNA CLCD is associated with the positive band in CD spectra in the area of absorption of nitrogen bases.

3.4 ACCUMULATION OF THE "GUEST" MOLECULES IN THE DNA MOLECULES FORMING CLCD PARTICLES

Figure 3.13, based on Figure 3.12, shows the structure of a quasinematic layer formed by double-stranded DNA molecules in a CLCD particle. The DNA molecules in the layer are located at an average distance, d, that depends on the osmotic pressure of

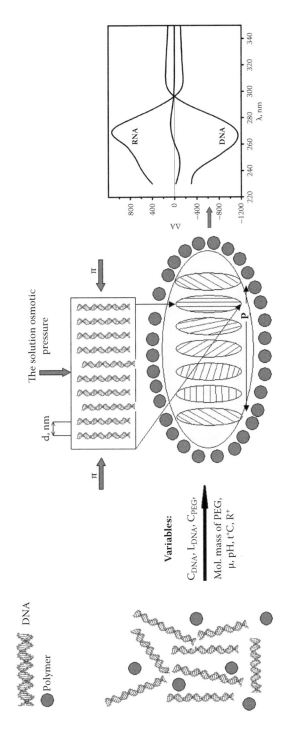

FIGURE 3.12 Scheme of the formation of CLCD particles at phase exclusion of DNA molecules from a water–salt–polymer-containing solution.

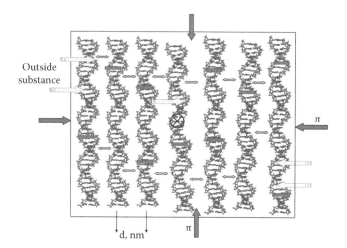

FIGURE 3.13 **(See color insert)** Structure of a quasinematic layer of DNA molecules in CLCD particles formed by phase exclusion of the molecules.

the solution (red arrows). The polymer is not included in the structure of the layer, and various compounds can easily diffuse into the layer and interact with the DNA molecules. The structure of a quasinematic layer explains the essence of the approach to the problem of accumulation of chemically or biologically active compound molecules (guest molecules) in CLCD particles.

First, this structure is retained by the osmotic pressure of the PEG-containing solution. Second, an ordered packing of adjacent DNA molecules is typical of CLCD particles [37]. Third, the liquid nature of DNA molecules packing in quasinematic layers provides the high rate of penetration of guest molecules into these layers and, hence, into CLCD particles. Fourth, the ordered location of DNA molecules in every layer does not limit the diffusion of guest molecules; moreover, it provides quick diffusion of guest molecules into a quasinematic layer. Finally, the retention of chemical reactivity of the structural elements of DNA molecules (nitrogen bases, etc.), their high local concentration and an ordered arrangement in the CLCD particles, which does not restrict the high rate of the diffusion of chemically relevant or biologically active compounds (guest molecules), can provide a high rate of their penetration into CLCD particles and reaction with DNA molecules. It is obvious that in the case of CLCD particles, guest molecules can be accumulated in the free space between the adjacent base pairs in DNA molecules. Indeed, linear double-stranded DNA molecules can carry such compounds as drugs or intercalators (i.e., planar compounds that can insert (or intercalate) between adjacent base pairs) in their structure. The equilibrium concentration of any of these compounds depends on the properties of the secondary structure of DNA (distance between the base pairs, their inclination, phosphate group charge density, etc.) and follows the known laws of adsorption.

Hence, various compounds can be chemically incorporated into the secondary structure of neighboring DNA molecules and, consequently, into the structure of the quasinematic layer and into the content of the CLCD particles.

The theory considered here predicts the appearance of an additional abnormal band in CD spectra if CLCD particles formed by double-stranded DNA molecules bind with compounds that, as well as the nitrogen bases, are rigidly fixed in respect to the long axis of the molecule (so-called external chromophores with an absorption band that does not coincide with the DNA nitrogen base absorption band). Meanwhile, at the low concentration of the external chromophores introduced into the structure of particles of DNA CLCD, the mode of packing of these molecules in the CLCD particles will not change. Such a situation is, indeed, possible, at the insertion of colored biologically active compounds (for instance, drugs) between the DNA base pairs. For this case, the theory is applied both to the DNA chromophores and the external chromophores introduced into the CLCD particle structure. It predicts the appearance of two abnormal bands (as a minimum!) located in different regions of the CD spectrum. One of these bands will still be in the DNA absorption region, while the other will appear in the region (for instance, visible region) where the external chromophores absorb. The sign of the abnormal band in the region of absorption of external chromophores (used drug) depends on their orientation in respect to the long axis of the DNA molecule. One can expect that if an angle of inclination of the drug molecule about the DNA axis is ≈90°, the sign of its abnormal band will coincide with that of the band typical of the DNA nitrogen bases. If the drug is located on the DNA molecule so that the angle of its inclination is within 0°–54°, the abnormal band in the CD spectrum can have a sign opposite to that of the band characteristic of the DNA nitrogen bases.

Figure 3.14 shows the CD spectrum theoretically calculated for the DNA CLCD treated with a colored drug, i.e., the anthracycline antibiotic daunomycin (DAU). (In Figure 3.14, D = 500 nm; P = 2,500 nm; curves 1–6 correspond to different

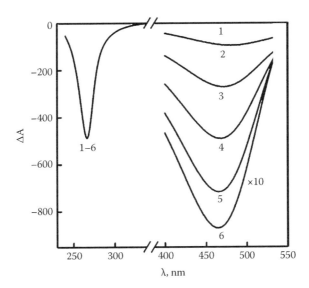

FIGURE 3.14 Theoretically calculated CD spectra for DNA CLCD particles treated with daunomycin.

concentrations of DAU.) One can see that the intercalation of the colored compound (external chromophore) into DNA molecules and, consequently, into the spatial structure of a CLCD particle, results in the appearance of two abnormal bands located in different regions of the CD spectrum. One of the bands is found in the absorption area of the DNA chromophores (nitrogen bases, $\lambda \approx 270$ nm), and the other lies in the absorption region of the drug molecules (external chromophores, $\lambda \approx 470$ nm). Both bands have negative signs at any extent of DAU binding to the DNA. The shapes of the bands in the CD spectra are identical to those of the absorption spectra for the DNA and the DAU.

The amplitude of the band in the CD spectrum in the region of DAU absorption rises with increasing number of its molecules bound to DNA, although the amplitude of the band in the region of DNA absorption remains constant. Note that the intercalation of DAU between the double-stranded DNA nitrogen base pairs does not interfere with the packing mode of neighboring DNA molecules in quasinematic layers, leaving the overall spatial structure of the DNA CLCD particles practically intact.

Figure 3.15 demonstrates the experimental CD spectra of the CLCD first formed by the double-stranded DNA molecules in PEG-containing solution (preformed CLCD) and then added DAU. In agreement with theoretical predictions, the CD spectra of the DNA CLCD colored by DAU have two bands. (In Figure 3.15, the values of C_{DAU} for curves 1–4 are 0, 1.2, 2.4, and 3.9×10^{-6} mol/L, respectively; C_{DNA} = 12 µg/mL; 0.3 mol/L NaCl; 170 mg/mL PEG; $\Delta A = (A_L - A_R) \times 1 \times 10^{-5}$ optical units.) One occurs in the absorption region of the DNA nitrogen bases ($\lambda \approx 270$ nm), and the other lies in the absorption region of DAU ($\lambda \approx 500$ nm). The shapes of the bands in the CD spectrum are identical to those in the absorption spectra of DNA and DAU. Under binding of DAU with double-stranded DNA molecules, both bands have negative signs despite DAU concentration.

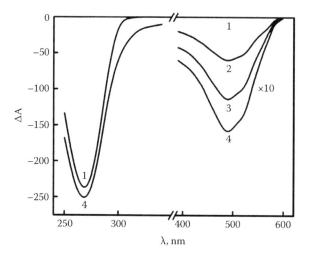

FIGURE 3.15 Experimentally measured CD spectra for the double-stranded DNA CLCD particles treated with daunomycin.

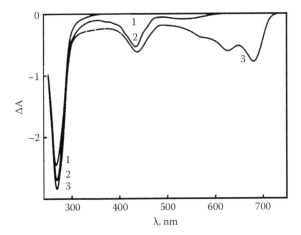

FIGURE 3.16 CD spectra of DNA CLCD in the absence and the presence of porphirine, as well as in the presence both porphine and mithoxanthrone.

The identical signs of two bands in the CD spectra of the double-stranded DNA CLCD colored by DAU simply means that the orientation of DAU molecules coincides with the orientation of the nitrogen base about the DNA axis. Hence, DAU molecules intercalate into DNA so that the angle between the DAU molecule and the long axes of the DNA is ≈90°. The appearance of two abnormal bands in different regions of the CD spectrum is evidence in favor of the cholesteric packing of DNA molecules in CLCD particles.

Figure 3.16 shows CD spectra of the CLCD particles formed by a DNA-porphine complex and the same particles treated with mitoxantrone (MX) as an example. (In Figure 3.16, porphine = 5.3×10^{-6} mol/L; $C_{MX} = 9.5 \times 10^{-6}$ mol/L; $C_{DNA} = 10$ μg/mL; $C_{PEG} = 170$ mg/mL; 0.3 mol/L NaCl; 0.002 mol/L phosphate buffer; pH 6.8; $\Delta A = (A_L - A_R) \times 2 \times 10^{-3}$ optical units.) One can see that in the first case there are two bands with identical signs in the CD spectrum. One of them is located in the absorption region of DNA nitrous bases ($\lambda \approx 270$ nm); the other one is in the absorption region of porphine chromophores. The shape of the bands in the CD spectrum both in the DNA absorption area and in the porphine absorption region corresponds to the shape of the absorption bands. The amplitude of the band in the porphine absorption area exceeds the amplitude typical of the molecular circular dichroism of this compound manifold. The coincidence of the signs of these bands located in different areas of the spectrum testifies that the porphine molecules intercalate between DNA base pairs. It can also be concluded from Figure 3.16 that the introduction of additional drug, i.e., MX, into the described system leads to the appearance of an additional band in the area of absorption of the antibiotic chromophores without affecting the location of the previous band. The result means that the structure of CLCD particles provides the conditions for the diffusion of various compounds and, consequently, compounds with different structural and optical parameters can be accumulated in DNA CLCD particles. Hence, one can consider every particle of the DNA CLCD as an effective adsorber for various compounds, and the process of immobilizing these

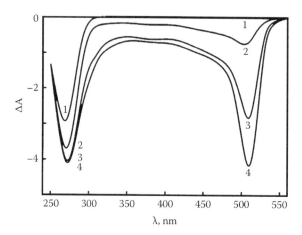

FIGURE 3.17   CD spectra of DNA CLCD in the absence and presence of SYBR Green.

compounds will induce an appearance of peculiarities (specific for each compound) in the CD spectra.

Recently, the colored cyanine drugs with very high extinction coefficients (for instance, compounds of SYBR Green group) have been synthesized. The CD spectra of DNA CLCD treated with SYBR Green are given in Figure 3.17. (Here curve 1 is C_t of SYBR Green = 0; curve 3 is $C_t = 9.8 \times 10^{-6}$; curve 4 is $C_t = 19.6 \times 10^{-6}$ mol/L of SYBR Green; $C_{DNA} = 10.2$ µg/mL; $C_{PEG} = 170$ mg/mL; 0.3 mol/L NaCl; 0.002 mol/L phosphate buffer; pH 6.8; $\Delta A = (A_L - A_R) \times 10^{-3}$ optical units.) One can see that the abnormal band still exists in the DNA absorption region, but an additional abnormal negative band in the SYBR Green absorption region that depends on its concentration is also apparent. Besides, in contrast to the cases of drugs considered previously, both bands have comparably high amplitudes. The coincidence of the signs of the abnormal bands in the absorption area of the DNA nitrogen bases and in the absorption region of the SYBR Green chromophores testifies that this compound intercalates between the DNA base pairs.

However, an interesting peculiarity of the SYBR Green is that this compound, when intercalating between base pairs of a linear double-stranded DNA molecule, begins to fluoresce [38], and the fluorescence is retained even in the case of the DNA CLCD particles formed in PEG-containing solutions. This is confirmed by results shown in Figure 3.18, where the fluorescence spectra of double-stranded DNA (curve 1) and DNA CLCD (curve 2), both treated with SYBR Green, are compared. (In Figure 3.18, $C_{DNA} \approx 8.89$ µg/mL; $C_{PEG} = 151.11$ mg/mL; $C_t = 9.8 \times 10^{-6}$ mol/L of SYBR Green; 0.3 mol/L NaCl + 10^{-2} mol/L phosphate buffer; pH ≈ 7.0.)

Moreover, new types of fluorescent microscopes have been developed. The combination of these technical possibilities opens a way to visualize the "image" of the DNA CLCD particles under the conditions shown here, where the structural parameters typical of these particles are not distorted.

Figure 3.19 shows the images of CLCD particles formed in a water–salt solution of PEG at different concentrations of DNA as an example. (In Figure 3.19, A depicts

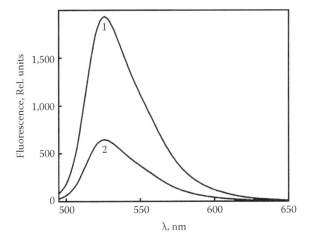

FIGURE 3.18   Fluorescence spectra of DNA and DNA CLCD in presence of SYBR Green.

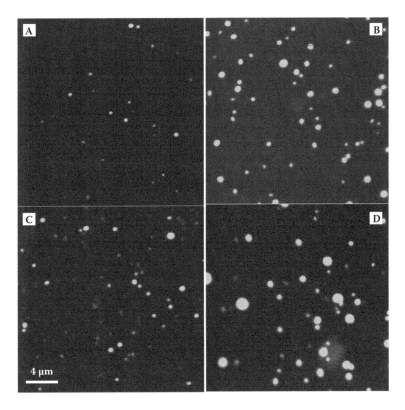

FIGURE 3.19   (See color insert) Fluorescence images of the DNA CLCD particles treated with SYBR Green.

C_{DNA} = 5 µg/mL, 10 minutes, C_t = 4.9 × 10^{-6} mol/L of SYBR Green; B depicts C_{DNA} = 5 µg/mL, 40 minutes, C_t = 4.9 × 10^{-6} mol/L of SYBR Green; C depicts C_{DNA} = 50 µg/mL, 10 minutes, C_t = 24.5 × 10^{-6} mol/L of SYBR Green; D depicts C_{DNA} = 50 µg/mL, 40 minutes, C_t = 24.5 × 10^{-6} mol/L of SYBR Green. C_{PEG} = 170 mg/mL; 0.3 mol/L NaCl; 0.002 mol/L phosphate buffer; pH 6.8. The scale bar is 4 µm. The images were taken by confocal microscope Leica TCS SP5.)

It can be seen that CLCD particles are existing as independent structural objects. Second, the average size of the CLCD particles depends on the DNA concentration used for the formation of the CLCD, which corresponds to the previous observations that the higher the concentration of DNA, the larger is the size of the CLCD particles [39, 40] formed by phase exclusion. Indeed, the fact that the average size of the particles formed in solutions with higher initial DNA concentration exceeds the size of the particles formed in solutions with lower DNA concentration (Figure 3.19, A and B) corresponds quite well with the theoretical evaluation discussed previously. The fluorescence images show that the size of the DNA CLCD particles formed in PEG-containing water–salt solution varies, depending on the DNA concentration, from 200 to 1,000 nm, which corresponds to the evaluation of the size of DNA dispersion particles discussed previously (at C_{DNA} → 0). The size of the DNA CLCD particles given previously is the result of a fine balance between the free energy of the particles and their surface free energy. The competition between the free energy of a dispersion particle (tends to increase the size of the particle) and its surface free energy (depends on the surface tension between the cholesteric phase and the isotropic phase and tends to reduce the surface between the DNA-rich phase and the isotropic solution surrounding it [41]) means that there is a critical size of a dispersion particle. Below this size, the particles are not stable or cannot be formed at all.

Finally, it can be noted that the size of the CLCD particles does not reach its optimal value immediately after the mixing of the DNA and PEG solutions, and the increase in the size to optimal value takes place within a certain time (Figure 3.19, A and B). This observation corresponds to the results of the analysis of the kinetics of the DNA CLCD particle formation. The kinetics of the process includes the appearance of an optically inactive liquid-crystalline "germ" and an increase in its size (i.e., the growth of a dispersion particle) followed by the twist of neighboring quasinematic layers of DNA molecules and the appearance of an abnormal optical activity of the CLCD particles [42, 43]. This kind of kinetics reflects the existence of an energetic barrier that the DNA molecules have to overcome so that the distribution of the DNA molecules between the isotropic phase outside the dispersion particles and the liquid-crystalline phase inside the particles can take place, making the twist of neighboring layers in the dispersion particles possible. The kinetics of the formation of DNA CLCD particles or synthetic polynucleotides is complicated and can be described within the framework of the classical Kolmogorov–Avrami equation [44] for the polymer crystallization.

As shown previously, a decrease in the PEG concentration that takes place at the dilution of PEG solutions containing DNA CLCD particles to minimal values is followed by the disintegration of the cholesteric structure of CLCD particles and the transition of the DNA molecules to an isotropic state. However, a certain period of

FIGURE 3.20 **(See color insert)** Images of DNA CLCD particles (green fluorescence) and cells transfected by recombinant plasmid expressing a red protein.

time is needed to realize this process. Taking into account this circumstance, there is an opportunity to observe the DNA CLCD particles treated with a fluorescent dye and placed directly into a cell culture.

Figure 3.20 compares the images of DNA CLCD (green fluorescence) and cells of human embryo kidney HEK-293T transfected with a recombinant plasmide pDS-Red2 that expresses a fluorescent red protein (REP). (Images were taken with Leica DMI-4000 B fluorescent microscope and kindly presented by Professor V. S. Prassolov.) One can see that the size of the formed DNA CLCD particles is much smaller in comparison to the size of the cells, which corresponds to the previously discussed evaluations of DNA CLCD particle size.

Therefore, these results show that liquid-crystalline DNA dispersions can easily accumulate molecules of various guests within their structure.

3.5 SUMMARY

In conclusion, one can say the following:

A. Linear, rigid, double-stranded DNA molecules are excluded from water–salt–polymer-containing solutions, and as a result, they form spatial liquid-crystalline dispersions. The size of particles of these dispersions can vary from 15 to 500 nm, depending on the length of the DNA molecules used.

B. The DNA molecules are arranged in quasinematic layers of their liquid-crystalline particles almost parallel to each other and, regardless of the fluctuations, tend to orient their long axes in one direction at any time. The local concentration of DNA in the particles of the cholesteric liquid-crystalline dispersion is very high and lies within 160–400 mg/mL, depending on the osmotic pressure of the solvent The distance between the DNA molecules packed in quasinematic layers of the spatial structure of dispersion particles can be changed from 30 to 50 nm, depending on the osmotic pressure of the solvent.

Points A and B demonstrate that the spatial parameters of the DNA liquid-crystalline particles correspond to the nanometer scale.

C. The rigid, anisotropic, double-stranded nucleic acid molecules in the liquid-crystalline particles realize their tendency to spatially twisted (cholesteric) packing, and the cholesteric mode of packing is retained until denaturation of the secondary structure of DNA (RNA) molecules. The cholesteric liquid-crystalline dispersions are characterized by an abnormal (intense) band in the CD spectrum.

Point C demonstrates that the optical properties of the DNA molecules in the liquid-crystalline dispersions differ from those typical of linear double-stranded DNA molecules.

Combination of points A–C allows one to suggest that the spatial structure formed by thousands of DNA molecules ordered in liquid-crystalline dispersion particles can be described by the term *DNA liquid-crystalline nanoconstruction* (LCNC) or simply *liquid DNA nanoconstruction*. (If a CLCD particle with an abnormal optical activity is formed, it can be called a DNA cholesteric liquid-crystalline nanoconstruction. The terms *DNA CLCD particle* and *DNA LCNC* are used as synonyms in the following chapters.)

Comparing the properties of DNA nanostructures (chapter 2) to the properties of DNA LCNC, it can be stressed that DNA LCNC consists of thousands of DNA molecules forming a *spatial* structure.

The unique peculiarity of DNA LCNC is the simplicity of manipulating the created structure. The simplicity and quickness of the formation of LCNC by double-stranded DNA molecules should also be noted, as well as the high yield (90%–95%) of the final product. The high local concentration of DNA and retention of the chemical reactivity of its components define the ability of DNA LCNC to absorb high concentrations of guest molecules. This point plays a key role in the issue of accumulation of guest molecules within DNA nanostructures and liquid DNA nanoconstructions.

Moreover, DNA LCNC properties can change flexibly under the influence of various factors, which makes the behavior of DNA LCNC similar to the behavior of biological objects such as the chromosomes of primitive organisms.

REFERENCES

1. Yevdokimov, Yu. M., S. G. Skuridin, and V. I. Salyanov. 1988. The liquid-crystalline phases of double-stranded nucleic acids in vitro and in vivo. *Liq. Crystals* 3:1443–59.
2. Yevdokimov, Yu. M., V. I. Salyanov, E. Gedig, and F. Spener. 1996. Formation of polymeric chelate bridges between double-stranded DNA molecules fixed in spatial structure of liquid-crystalline dispersions. *FEBS Lett.* 392:269–73.
3. Seeman, N. C. 2007. An overview of structural DNA nanotechnology. *Mol. Biotechnol.* 37:246–57.
4. Douglas, S. M., J. J. Chou, and W. M. Shih. 2007. DNA-nanotube-induced alignment of membrane proteins for NMR structure determinations. *Proc. Natl. Acad. Sci. USA* 104:6644–48.
5. Andersen, E. S., M. Dong, M. M. Nielsen, et al. 2009. Self-assembly of nanoscale DNA box with a controllable lid. *Nature* 459:73–76.
6. LaBean, T. H. 2009. Another dimension for DNA art. *Nature* 459:331–32.
7. Douglas, S. M., H. Dietz, T. Liedl, B. Hogberg, F. Graf, and W. Shih. 2009. Self-assembly of DNA into nanoscale three-dimensional shapes. *Nature* 459:414–18.
8. Cheng, W. L., M. J. Campolongo, J. J. Cha, et al. 2009. Free-standing nanoparticle superlattice sheets controlled by DNA. *Nature Materials* 8:519–25.
9. Yevdokimov, Yu. M., V. I. Salyanov, S. V. Semenov, and S. G. Skuridin. 2011. *DNA Liquid-Crystalline Dispersions and Nanoconstructions*, ed. Yu. M. Yevdokimov. Boca Raton, FL: Taylor & Francis.
10. Adamczyk, A. 1989. Phase transition in freely suspended smectic droplets: Cotton-Mouton technique, architecture of droplets and formation of nematoids. *Mol. Cryst. Liq. Cryst.* 170:53–69.
11. Yevdokimov, Yu. M., S. G. Skuridin, and G. B. Lortkipanidze. 1992. Liquid-crystalline dispersions of nucleic acids. *Liq. Crystals* 12:1–16.
12. Yevdokimov, Yu. M. 1991. Liquid-crystalline dispersions of nucleic acids. *Bulletin of the USSR Academy of Sciences, a physical series* (Russian ed.) 55:1804–16.
13. Salyanov, V. I., V. G. Pogrebnyak, S. G. Skuridin, et al. 1978. On the relation between the molecular organization of the solution of poly(ethylene glycol)-water and the compactization of the double-stranded DNA molecules. *Molecular Biology* (Russian ed.) 12:485–95.
14. Minagava, K., Y. Matsuzawa, K. Yoshikawa, A. R. Khokhlov, and M. Doi. 1994. Direct observation of coil-globule transition in DNA molecules. *Biopolymers* 34:555–58.
15. Tanaka, F. 1983. Concentration-dependent collapse of a large polymer in a solution of incompatible polymers. *J. Chem. Phys.* 78:2788–94.
16. Evdokimov, Yu. M., N. M. Akimenko, N. E. Glukhova, and Ya. M. Varshavsky. 1974. DNA compact form in solution. I: Patterns of absorption spectra of polyribonucleotides and DNA in PEG-containing water-salt solutions. *Molecular Biology* (Russian ed.) 8:396–405.
17. Grosberg, A. Yu., I. Ya. Erukhimovich, and E. I. Shakhnovich. 1981. On DNA compactization in diluted polymeric solutions. *Biophysics* (Russian ed.) 26:897–905.
18. Belyakov, V. A., and A. S. Sonin. 1982. *Optics of Cholesteric Liquid Crystals* (Russian ed.). Moscow: Nauka.
19. Sonin, A. S. 1983. *Introduction in Physics of Liquid Crystals* (Russian ed.). Moscow: Nauka.
20. Livolant, F., and M. F. Maestre. 1988. Circular dichroism microscopy of compact forms of DNA and chromatin in vivo and in vitro: Cholesteric liquid crystalline phases of DNA and single dinoflagellate nuclei. *Biochemistry* 27:3056–68.
21. De Vries, H. 1951. Rotatory power and other optical properties of certain liquid crystals. *Acta Cryst.* 4:219–26.

22. De Gennes, P.-G. 1974. *The Physics of Liquid Crystals.* London: Oxford University Press.

23. Saeva, F. D., P. E. Sharpe, and G. R. Olin. 1973. Cholesteric liquid crystal induced circular dichroism (LCID). V: Mechanic aspects of LCID. *J. Amer. Chem. Soc.* 95:7656–59.

24. Maestre, M. F., and C. Reich. 1980. Contribution of light scattering to the circular dichroism of deoxyribonucleic acid films, deoxyribonucleic acid-polylysine complexes, and deoxyribonucleic acid particles in ethanolic buffers. *Biochemistry* 19:5214–23.

25. Keller, D., and C. Bustamante. 1986. Theory of the interaction of light with large inhomogeneous molecular aggregates. I: Absorption. *J. Chem. Phys.* 84:2961–71.

26. Keller, D., and C. Bustamante. 1986. Theory of the interaction of light with large inhomogeneous molecular aggregates. II: Psi-type dichroism. *J. Chem. Phys.* 84:2972–80.

27. Kim, M.-H., L. Ulibarri, D. Keller, and C. Bustamante. 1986. The psi-type dichroism of large molecular aggregates, III: Calculations. *J. Chem. Phys.* 84:2981–89.

28. Purcell, E. M., and C. R. Pennypacker. 1973. Scattering and absorption of light by nonspherical dielectric grains. *Astrophys. J.* 186:705–14.

29. Drain, B. T. 1988. The discrete-dipole approximation and its application to interstellar graphite grains. *Astrophys. J.* 333:848–72.

30. Belyakov, V. A., and V. E. Dmitrienko. 1989. Optics of chiral liquid crystals. *Sov. Sci. Rev. A Phys.* 13:1–212.

31. Belyakov, V. A., E. I. Demikhov, V. E. Dmitrienko, and V. K. Dolganov. 1985. Optical activity, transmission spectra, and structure of blue phases of liquid crystals. *JETP* (Russian ed.) 89:2035–51.

32. Belyakov, V. A., S. M. Osadchii, and V. A. Korotkov. 1986. Optics of imperfect cholesteric liquid crystals. *Crystallography Reports* (Russian ed.) 31:522–27.

33. Belyakov, V. A., V. P. Orlov, S. V. Semenov, et al. 1996. Some features of circular dichroism spectra of liquid-crystalline dispersions of double-stranded DNA molecules and DNA complexes with strained compounds. *Biophysics* (Russian ed.) 41:1044–55.

34. Grasso, D., S. Fasone, C. La Rosa, and V. Salyanov. 1991. A calorimetric study of the different thermal behaviour of DNA in the isotropic and liquid-crystalline states. *Liq. Crystals* 9:299–305.

35. Evdokimov, Yu. M., T. L. Pyatigorskaya, N. A. Belozerskaya, et al. 1977. DNA compact form in solution. XI: Melting of the DNA compact state, formed in water-salt solutions, containing poly(ethylene glycol). *Molecular Biology* (Russian ed.) 11:507–15.

36. Grasso, D., R. G. Campisi, and C. La Rosa. 1992. Microcalorimetric measurements of thermal denaturation and renaturation processes of salmon sperm DNA in gel and liquid crystalline phases. *Thermochim. Acta* 199:239–45.

37. Goldar, A., H. Thomson, and J. M. Seddon. 2008. Structure of DNA cholesteric spherulitic droplet dispersions. *J. Phys. Condens. Matter* 20:1–9.

38. Zipper, H., H. Brunner, J. Bernhagen, and F. Vitzthum. 2004. Investigation on DNA intercalation and surface binding by SYBR Green I, its structure determination and methodological implications. *Nucl. Acids Res.* 32:e103.

39. Yevdokimov, Yu. M., S. G. Skuridin, and N. M. Akimenko. 1984. Liquid-crystalline microphases of low-molecular double-stranded nucleic acids and synthetic polynucleotides. *Polymer Science, series A* (Russian ed.) 26:2403–10.

40. Grosberg, A. Yu., and A. R. Khokhlov. 1994. *Statistical Physics of Macromolecules.* New York: American Institute of Physics.

41. Ubbink, J., and T. Odijk. 1995. Polymer- and salt-induced toroids of hexagonal DNA. *Biophys. J.* 68:54–61.

42. Skuridin, S. G., E. V. Shtykova, and Yu. M. Evdokimov. 1984. Formation kinetics of optically active liquid-crystalline microphases of low molecular DNA. *Biophysics* (Russian ed.) 29:337–38.

43. Skuridin, S. G., G. B. Lortkipanidze, O. R. Musaev, and Yu. M. Yevdokimov. 1985. Formation of liquid-crystalline microphases of two-chain nucleic acids and synthetic polynucleotides of low molecular mass. *Polymer Science, series A* (Russian ed.) 27:2266–73.
44. Wunderlich, B. 1979. *Macromolecular physics*. Vol. 2 (Russian ed.). Moscow: Mir.

4 "Rigid" Nanoconstructions Based by Spatially Ordered Double-Stranded DNA Molecules Complexed with Various Compounds and Nanoparticles

4.1 BACKGROUNDS FOR FORMATION OF RIGID DNA NANOCONSTRUCTIONS

In the system used for the double-stranded DNA CLCD (cholesteric liquid-crystalline dispersions) formation, the packing mode of DNA molecules in the particles of dispersion is determined at the "moment of their recognition" during the approach of the DNA molecules, which have phosphate groups neutralized by counterions. After formation of the particles of DNA CLCD, the mutual orientation of the DNA in these particles is constrained, although DNA molecules retain some diffusion degrees of freedom in quasinematic layers. This means that the osmotic pressure of the PEG solution, determined by PEG concentration, controls the packing mode of neighboring, linear, rigid, native DNA molecules in particles [1, 2]. The constant osmotic pressure of the solution determines not only a constant spatial structure of the DNA CLCD particles and, consequently, a fixed distance between DNA molecules in the quasinematic layers, but a fixed value of the amplitude of an abnormal band in the CD spectrum in the DNA absorption region.

These results that specify the properties of the LCDs formed by linear, rigid, double-stranded DNA molecules with low molecular mass (or other double-stranded polynucleotides) as well as peculiarities of their CD spectra allow one to define a few points that have an important meaning for creation of "rigid" nanoconstructions based on DNA molecules fixed in the content of LCNC (liquid-crystalline nanoconstruction).

Figure 3.13, where the hypothetical structure of a so-called quasinematic layer typical of a double-stranded DNA LCNC (or DNA CLCD) was shown, explains the essence of our approach to designing DNA-based rigid nanoconstructions. This figure focuses attention on a number of facts necessary for creation of these nanoobjects, i.e., objects with properties that are principally different from those of both the "liquid" DNA nanoconstruction and the DNA nanostructures (see chapter 2).

1. At a fixed osmotic pressure of the solution, the mutual orientation of neigh-boring double-stranded DNA molecules in LCNC is "frozen," although some diffusion degrees of freedom of the DNA molecules are retained to determine the liquid mode of these molecules packing.

2. The distance between neighboring DNA molecules in a layer can be regu-lated from 3.0 to 5.0 nm only by a change in the osmotic pressure of the PEG solution or (at a fixed concentration and molecular mass of PEG) by a change in the temperature of the solution.

3. There is enough free space between DNA molecules ordered in a layer; at the same time, PEG is not included into the composition of the LCNC.

4. Given the significant distance between DNA molecules, the liquid mode of packing of the molecules and their high concentration provide the condi-tions for quick diffusion of the molecules of many compounds ("guest" molecules) both between DNA molecules in one layer and between the DNA molecules in adjacent layers of LCNC.

5. The interaction of the chemical compounds with the reactive groups of DNA molecules (at certain conditions) does not distort the mode of DNA molecules packing in a layer, but it can bring a new chemical reactivity to the whole structure.

6. From a theoretical point of view, there are no limitations to the arrangement of guest molecules in the free space between DNA molecules in a layer. In particular, one can try to reach a high concentration of the guests by includ-ing them into cross-links between all of the neighboring DNA molecules.

7. For the formation of the cross-links, it is necessary that there be sites on the surface of the DNA molecules where the cross-links can be terminated.

These points raise an interesting question: Is it possible to transform the liquid structure of this layer to its rigid state so that (a) the spatial structure of the particles does not change and (b) the created structure would be able to exist in the absence of the osmotic pressure of PEG solution?

The theoretical description of the basic properties of DNA CLCDs allows one to suggest that there are two possible ways to answer this question:

1. By cross-linking neighboring DNA molecules located both in the same and neighboring quasinematic layers in the particles of the initial LCNC
2. By increasing the interaction between double-stranded DNA molecules (or their fragments) by the action of a chemical compound on these molecules that would lead to a decrease in the solubility of the DNA molecules and the transition of LCNC into an insoluble (rigid) state

The first answer initiates two additional questions.

1. How does one link the DNA molecules with cross-links?
2. What will be the properties of the LCNC containing the cross-links?

The fundamental problem that needs to be solved can be formulated as follows: Using the particles of the initial DNA CLCD, the adjacent DNA molecules in CLCD particles have to be linked such that the spatial ordering of DNA molecules in CLCD particles is not distorted.

4.2 RIGID NANOCONSTRUCTIONS FORMED AS A RESULT OF FORMATION OF NANOBRIDGES BETWEEN NEIGHBORING DNA MOLECULES FIXED IN QUASINEMATIC LAYERS

From a theoretical point of view, it is obvious that in order to form the cross-links between both neighboring DNA molecules in the same quasinematic layer (Figure 3.13) and between DNA molecules in neighboring layers, these molecules should have the terminal sites for cross-link formation. This process requires, first, the presence of terminal sites located mainly on the surface of the double-stranded DNA molecules. Theoretically, such a terminal site can be metal ions specifically fixed in a groove (or grooves) on the DNA surface capable both of forming complexes with the DNA nitrogen bases and of attaching other ligands that can be introduced into the system. In addition, one can use as terminal sites the reactive groups in the content of planar compounds, which form various complexes with DNA molecules. Second, taking into account the steric location of the DNA reactive groups (in particular, the nitrogen ^7N atoms of purines) in the grooves on the surface of the double-stranded DNA molecules, it is evident that two neighboring DNA molecules can be cross-linked only if the spatial orientations of these groups are coordinated (i.e., the neighboring DNA molecules must be sterically "phased" to realize the cross-linking of the same reactive groups that belong to various DNA molecules). Finally, the theory predicts that when the cross-links between DNA molecules contain chromophores and these cross-links are specifically oriented with respect to the long axes of DNA molecules in quasinematic layers, one can expect an appearance of an intense band in the CD spectrum located in absorption region of the chromophores. All of these points demonstrate that the formation of cross-links is a delicate process that requires a number of conditions.

Hence, the problem is reduced to the formation of artificial cross-links with adjustable properties between neighboring DNA molecules fixed (due to constant osmotic pressure of the PEG-containing solution) at an interhelical distance of about 3.5 nm in quasinematic layers in the structure of LCNC (Figure 3.13) without significant change in the total spatial organization of this nanoobject. (Taking into account the length of possible cross-links, one can consider them "nanobridges" between DNA molecules.) The nanobridges must link the DNA molecules located both in one layer and in neighboring layers of the liquid-crystalline structure, which leads to the "freezing" of all of the diffusion degrees of freedom of the DNA molecules. This, in turn, will cause a dramatic change in the properties of the resultant structure. In particular, the liquid mode of packing of adjacent DNA molecules and the diffusive mobility of the molecules may disappear, and the structure itself may obtain the properties of a rigid (solid) material. Finally, if the cross-linking process is realized so that the total spatial structure of LCNC is retained, the abnormal optical activity will allow one to control changes both in the secondary structure of the initial DNA molecules and the fixation of the chemical components in the content of nanobridges.

It is evident, that the efficiency of the nanobridge formation will depend on the parameters of the available free space between DNA molecules as well as the number of chemical components in the content of nanobridges. However, the efficiency apparently is not directly connected to the peculiarities of the secondary structure of double-stranded nucleic acid molecules. This method was first formulated in 1996 [3], but results were obtained only recently through collaboration of scientific teams from Russia, Italy, and Germany [4–7].

The detailed analysis of the properties of compounds that can be used as components for extended nanobridges made it possible to choose the molecules of the anthraquinone (anthracycline) group. These compounds, owing to their chemical structure, can form chelate complexes with ions of bivalent metals. The chelate complexes with bivalent copper ions attract special interest because of the specific electron structure of a bivalent copper ion [8] and the spatial structure of the anthraquinones, which form chelate complexes that have a planar (flat) structure. The structure of a polymeric chelate complex is shown in Figure 4.1(A). These chelate complexes may contain up to ten repeating subunits; that is, under certain conditions,

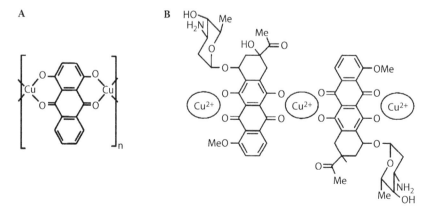

FIGURE 4.1   Chelate complexes of anthraquinones (A) and anthracyclines (B) with bivalent copper.

they can form extended planar structure with copper ions [9, 10] (so-called polymer chelate complexes [11] with copper ions). At the same time, a chelate complex can start and terminate with both an anthraquinone molecule and a bivalent copper ion (Figure 4.1).

The water-soluble antibiotics of the anthracycline group form flat, rigid, chelate complexes with metal ions (Figure 4.1(B)). The formation of chelate complexes with bivalent copper ions is the most effective. The molecules of this group of antibiotics, namely, daunomycin (DAU), can intercalate between the base pairs of DNA and synthetic polynucleotides of the B-family only, i.e., they can form an intercalation complex. At the intercalation, the reactive groups of the DAU molecules (keto-oxygen, peri-OH-groups) are not available for chemical reactions. However, DAU and its analogues can form so-called external complexes with molecules of nucleic acids of B- and A-families [12, 13]. At the formation of an external complex, the DAU reactive groups become available for chemical reactions, namely, for chelate formation.

Analysis of the various anthracyclines that differ by the presence and the location of aglycone substituents has shown that one of the necessary conditions at the formation of extended flat complexes is the presence of reactive oxygen atoms at the positions 5, 6 and 11, 12 of anthracycline aglycone.

The details of the technology of treatment of DNA CLCD with a solution of DAU and a copper salt are thoroughly described in the literature [4, 5, 14]. Formation of cross-links (nanobridges) was realized according to the following scheme: Initially, the DNA CLCD was obtained by mixing equal volumes of water–salt DNA and PEG solutions. Fixed volumes of daunomycin (DAU) solution were subsequently added to the DNA CLCD. Finally, this mixture was treated with a small volume of $CuCl_2$ solution while stirring.

Figure 4.2 compares the CD spectra of all structures formed by DNA molecules. (In Figure 4.2, curve 1 depicts the CD spectrum of the initial DNA CLCD; curve 2 depicts the CD spectrum of DNA CLCD after DAU processing; and curve 3 depicts the CD spectrum of [DNA-DAU] complex CLCD after $CuCl_2$ treatment. Conditions:

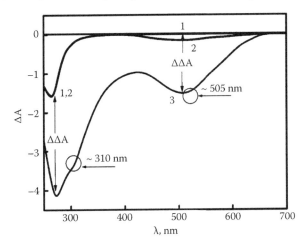

FIGURE 4.2 CD spectra of the initial DNA CLCD and the same dispersion treated with daunomycin and $CuCl_2$.

$C_{DNA} = 5.5$ µg/mL, $C_{PEG} = 170$ mg/mL, 0.3 mol/L NaCl, 0.002 mol/L phosphate buffer, pH 6.8, $\Delta A = (A_L - A_R) \times 10^{-3}$ optical units, $\Delta\Delta A$ shows the increase in the amplitude of an abnormal band as a result of nanobridge formation.)

The formed dispersion was treated with DAU (curve 2) and then by CuCl$_2$ solution (curve 3). The amplitude of the intense negative band at $\lambda \approx 270$ nm indicates the left-handed twist of the spatial structure of CLCD and remains practically unchanged at any reasonable DAU concentration added to the CLCD. An addition of DAU to the DNA CLCD is accompanied by an appearance of a new band located in the absorption region of DAU chromophores ($\lambda \approx 500$ nm). The amplitude of this band grows with increasing DAU concentration, reaching an equilibrium value at the DAU concentration that corresponds to the maximal extent of the DNA saturation by DAU molecules and does not change upon further growth of DAU concentration. (Note that the equilibrium amplitude of the band in the CD spectrum characteristic of [linear DNA-DAU] complex (not liquid-crystalline!) does not exceed a few units of ΔA.) The *negative* sign of the band at $\lambda \approx 500$ nm reflects an intercalation of DAU molecules between the nitrogen base pairs of DNA molecules, fixed at a particular distance due to the osmotic pressure of a PEG solution.

The addition of CuCl$_2$ solution to the DNA CLCD treated by DAU with an equilibrium band amplitude of $\lambda \approx 500$ nm results in a manyfold increase (amplification) of both this band and the band located in the UV region of the spectrum (curve 3). For instance, under the conditions used (DNA molecular mass $= 8 \times 10^5$ g/mol and DNA concentration ≈ 5 µg/mL), the maximum amplitude of the band at $\lambda \approx 500$ nm is very high and equals approximately -2500 units (!) of ΔA. The amplification of both of these bands begins only after achieving a critical concentration of copper ions in solution.

It should be noted that no substantial alterations in the CD spectrum in the DAU absorption region were observed when cholesteric LCD composed of double-stranded RNA molecules (A-family) were treated with DAU. The lack of a strong CD band in the region of DAU absorption indicates that no intercalation complex is formed between DAU and RNA. Hence, in contrast to DNA, DAU molecules are located isotropically near the RNA surface, i.e., they can form only an external complex, and these molecules are practically invisible in the CD spectrum.

However, an addition of CuCl$_2$ solution to the DNA CLCD treated with DAU and having an equilibrium value of the amplitude of the band at $\lambda \approx 500$ nm results not only in a manyfold increase (amplification) of this band, but also a band located in the UV region of the spectrum (curve 3). (The bands at $\lambda \approx 500$ nm and $\lambda \approx 300$ nm are characteristic of the CD spectrum of linear, isotropic [DAU-Cu^{2+}] complexes; this reflects the existence of low-energy ($\lambda \approx 500$ nm) and high-energy ($\lambda \approx 300$ nm) electronic transitions in DAU moieties of the complexes. These two bands are maintained at the formation of a complex between the linear DNA and DAU.) Indeed, upon addition of CuCl$_2$, amplification of the CD spectrum was observed along with a change in the shape of the band in the UV region (curve 3). In particular, curve 3 in Figure 4.2 has a shoulder at $\lambda \approx 310$ nm.

Most important is that similar changes took place in the CD spectrum of RNA CLCD or double-stranded poly(A)×poly(U) and triple-stranded polyribonucleotide poly(A)×2poly(U) molecules (A-family) subsequently treated by DAU and Cu^{2+} ions, where intercalation of DAU is not possible at all due to steric reasons. Therefore, for all the investigated CLCD particles formed by nucleic acids and synthetic polynucleotides of B- or A-families, after their treatment with DAU and bivalent copper ions, a significant amplification of optical activity of CLCD in the area of the DAU chromophore absorption region is observed.

Here, we stress again that DAU molecules located isotropically along neighboring nucleic acid molecules are practically invisible in the CD spectrum of the CLCD. In addition, anisotropic location of DAU molecules, due to their intercalation between DNA nitrogen base pairs, does not induce the extra increase of the band in the CD spectrum after addition of $CuCl_2$, because complex formation is not possible.

Hence, one can infer that there is a quite different mechanism that explains the extra increase of the CD band after addition of $CuCl_2$. This mechanism does not need intercalation of DAU molecules between nitrogen base pairs. The amplification of the bands of the DNA CLCD, as well as RNA CLCD, in the visible and UV regions shows that, under the conditions used, a portion of the DAU molecules is indeed located near the DNA (or RNA), forming external (nonintercalative) complexes. The DAU molecules of the external complexes are acceptable for chemical reaction, for instance, for formation of a chelate complex with Cu^{2+} ions.

Because amplification of the optical properties of isotropically located chromophores is impossible, the results presented here suggest that the [DAU-Cu^{2+}] complexes should be ordered near the surface of nucleic acid molecules in the content of the particles of cholesteric LCDs. Indeed, according to the theory, the amplification of optical properties is possible only for the chromophores that are severely fixed in respect to the "director" of the quasinematic layer of the cholesteric liquid crystal.

The manyfold amplification of the bands at $\lambda \approx 505$ and ≈ 310 nm indicates the appearance of an additional (as well as intercalation) type of anisotropic arrangement of DAU molecules in proximity to DNA molecules. The anisotropic arrangement of [DAU-Cu^{2+}] complexes, which differs from common intercalation of DAU and causes the amplification of the 505- and 310-nm bands in the CD spectrum of the DNA CLCD, could be explained by two different rationales. First, one may suppose that, owing to stacking interaction between DAU molecules, this is caused by the formation of vertical stacks (n-mers) of DAU molecules near the DNA surface in the structure of the CLCD particles. This means that a shell of DAU appears in proximity to the surface of a DNA molecule where a portion of the DAU molecules are bridged by Cu^{2+} ions. It is obvious that the direction of the vertical axis of the resulting structure of DAU n-mers coincides with the direction of the DNA long axis. Second, one may suppose that complexes [DAU-Cu^{2+}] are located between the neighboring NA molecules in such a way that they form cross-links between these molecules. It should be noted that, in principle, the beginning of the cross-link could be formed not only by a DAU molecule forming an external complex with nucleic acid molecules, but also by Cu^{2+} ions chelating the nitrogen bases. The direction of the long axis of the cross-links, formed by [DAU-Cu^{2+}] complexes, proves to be perpendicular to the direction of the long axis of the nucleic acids, although the orientation of DAU molecules is close to that of the nitrogen base pairs.

The properties of the DNA CLCD particles listed here allow choosing between variants in the disposition of the [DAU-Cu^{2+}] complexes. In the case of the first assumption, the main factor stabilizing the liquid-crystalline structure of DNA particles, even covered with DAU molecules, is still the osmotic pressure of the aqueous polymer solution. Consequently, violation of the boundary conditions necessary for the formation of dispersions, in particular dilution of water–salt–polymer-containing solution, which is accompanied by a decrease in the osmotic pressure, should result in the transition of DNA molecules from liquid-crystalline into isotropic state. Isotropic DNA solution is known to be devoid of abnormal optical activity. This means that, in the first case, the dilution should result in the disintegration of the cholesteric structure of LCD and in the disappearance of abnormal optical activity. In the case of the second assumption, the osmotic pressure of solution is not the main factor affecting the character of packing DNA molecules in a CLCD particle. In this case, the abnormal optical activity should persist even upon dilution, and the specific optical activity should remain constant.

It was shown that multiple dilutions of PEG-containing water–salt solution do not lead to an appreciable decrease in the specific abnormal optical activity of the DNA CLCD treated by DAU and $CuCl_2$, meaning that the mutual orientation of neighboring DNA molecules is not violated even outside the boundary conditions. This is possible only if neighboring DNA molecules are *indeed cross-linked*, and cross-links stabilize the cholesteric structure of the CLCD particles. Thus, the reason for the increase in the amplitude of bands located in different regions of the CD spectrum of CLCD of DNA successively treated by DAU and $CuCl_2$ is the formation of cross-links between neighboring DNA molecules.

The formation of cross-links between neighboring DNA molecules suggests an interesting correlation between the abnormal optical activity of DNA CLCD and the distance between DNA molecules. Indeed, it is known that the condensation of the DNA molecules in PEG-containing solutions is not accompanied by an alteration in the parameters of their secondary structure (B-family) or in the chemical reactivity of nitrogen bases, and, hence, the abilities of these molecules to interact with DAU and $CuCl_2$.

It means that possible chemical sites for the termination of cross-links are constant and not affected by the change in PEG concentration. However, the increase in PEG concentration results in a decrease in the distance between neighboring DNA molecules (Figure 3.11). Hence, upon formation of cross-links located *between* neighboring DNAs, the length of these cross-links will depends on the distance between these molecules. Because in this case the cross-links are fixed perpendicularly in respect to the direction of the long axes of the DNA molecules and they consist of $[DAU-Cu^{2+}]$ complexes, one can expect that the greater the distance between neighboring DNA molecules fixed in the structure of the CLCD particles, the higher will be the length of cross-links anisotropically fixed between DNA molecules.

Figure 4.3 compares the experimentally measured dependencies of the maximal amplitude of the CD band ($\lambda = 500$ nm) of DNA CLCD treated with DAU and $CuCl_2$ and the distance between DNA molecules as a function of PEG concentration. (In Figure 4.3, curve 2 represents the effect of PEG concentration on distance, D, between DNA molecules based on the value of small-angle reflection in an X-ray diagram of DNA liquid crystals, where $C_{DNA} = 5$ µg/mL; curve 1 was obtained in the presence of $C_{DAU} = 13.9 \times 10^{-6}$ mol/L and 2×10^{-5} mol/L $CuCl_2$, taking into account

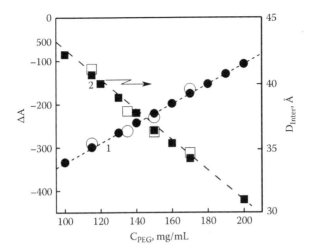

FIGURE 4.3 Change in the amplitude of the CD band of DNA CLCD added with dauno-mycin and CuCl$_2$ and in the distance between DNA molecules in CLCD particles upon PEG concentration.

that all DNA molecules are included in CLCD particles. Conditions: 0.3 mol/L NaCl + 0.002 mol/L phosphate buffer; pH 6.67; ΔA_{500} is given in mm; 1 mm = 10^{-5} optical units.) One can see that an increase in PEG concentration is accompanied by a decrease in the average distance between the axes of the neighboring DNA molecules fixed in CLCD particles (curve 2). Moreover, the lower the PEG concentration, the higher the absolute increase in the CD band of CLCD formed by DNA molecules containing cross-links. Figure 4.3 demonstrates that an extra increase in the amplitude of the CD band is connected with an increase in the total length of [DAU-Cu^{2+}] complexes, i.e., chromophores anisotropically fixed in respect to the long axes of the DNA molecules. The greater the distance between neighboring DNA molecules in CLCD particles, the higher the abnormal optical activity of these particles treated with DAU and CuCl$_2$. This is possible only in the case of the formation of cross-links consisting of [DAU-Cu^{2+}] chelate complexes *between* neighboring DNA molecules. There are also other ways to prove the formation of cross-links between DNA molecules fixed in CLCD particles [4, 5, 14].

Due to the presence of the cross-links (or nanobridges) between neighboring DNA molecules, the spatial structure of CLCD particles is retained. Such structure remains stable for a long time (months) at PEG and NaCl concentrations significantly below those at which the CLCD particles, not stabilized with nanobridges, can exist [15]. Consequently, the created structure can exist not only in solutions with low ionic strength, but also in solutions with a very low (up to zero) concentration of PEG, i.e., under the conditions of a low osmotic pressure of the solution.

The efficiency of the formation of nanobridges depends on concentrations of both DAU molecules and copper [14]. Both of the dependencies are S-shaped with a high extent of consistency. Amplification of the band begins only after the achievement of a critical concentration (C^{cr}) of both copper ions and DAU molecules in the solution.

For DAU molecules, the critical concentration means that those DAU molecules that arise after the DAU intercalation take part in the formation of nanobridges. The critical concentration for copper ions means that these ions can induce some changes in the DNA secondary structure, after which copper ions (or their complexes with base pairs and DAU molecules) become available for the further chelation that is necessary for nanobridge formation. (This, in turn, means that the order of the addition of components, i.e., DAU and copper ions, is important for the creation of nanobridges.) The close similarity of the type of these dependencies reflecting the amplification of the bands located in different regions of the CD spectrum makes it possible to assume that the formation of the nanobridges somehow initiates the transition between different structural forms of CLCD particles, namely, the transition between the initial liquid structure and a new structure that evolves after the DNA molecules are linked with the nanobridges.

4.3 THE MAGNETOMETRIC EVALUATION OF CU²⁺ IONS IN THE CONTENT OF NANOBRIDGES FORMED BETWEEN SPATIALLY FIXED DNA MOLECULES

The calculation of the number of Cu^{2+} ions in the content of the nanobridges is based on the measurement of the magnetic susceptibility (magnetic moment) of a Cu^{2+} ion being in the d^9 state with four reactive oxygen atoms at the nanobridge formation [16].

The low-temperature changes in the magnetic susceptibility of the DNA CLCD treated with DAU and $CuCl_2$, i.e., the samples containing the nanobridges, were measured at a magnetic field of 71.29 mT (712.9 Oe) by the superconducting interferometer device (SQUID-magnetometer) produced by D. I. Mendeleev University in Moscow [17].

The number of quanta (N) of the magnetic flow through the sample was measured. This quantity is proportional to the magnetic moment (P_m) of the explored sample. In this case, the paramagnetic centers are only the Cu^{2+} ions.

Using the experimentally measured temperature dependence of the magnetic moment upon temperature, the magnetic susceptibility was calculated:

$$\chi = P_m/H \tag{4.1}$$

where H is the magnetic field value that is equal to 712.9 Oe.

The dependence of $\chi(T)$ is well approximated by the well-known analytical dependence:

$$\chi = \chi_0 + C/(T - T_C) \tag{4.2}$$

where $C = (\mu_{eff})^2 \times N^*/(3k)$ and N^* is the number of the paramagnetic centers of Cu (d^9).

Using the data of Nikiforov et al. [17], some practically important parameters have been calculated. One can estimate an effective magnetic moment of one Cu^{2+} ion in the DNA CLCD sample as:

$$\mu_{eff} = g\mu_B \sqrt{j\,(j+1)} = 1.82\;\mu_B \qquad (4.3)$$

From temperature dependence of magnetic moment, it is possible to evaluate the number of Cu^{2+} ions in the d^9-state, N, as:

$$N = 3kC(\mu_{eff})^2 = 1.96 \times 10^{18} \qquad (4.4)$$

where $C = 1.47 \times 10^{-6}$ (CGS units) is the constant obtained from an approximation of the temperature dependence of magnetic susceptibility according to the Curie–Weiss law.

As the number of DNA molecules in the sample $N_{DNA} = 2.98735 \times 10^{15}$, there are approximately $N*d^9/N_{DNA} = 716$ Cu^{2+} ions per one DNA molecule. As the number of helical turns in the DNA molecule with molecular mass of 8×10^5 g/mol is equal to 120, from here it follows that on each turn of DNA helix, $716/120 = 5.9$ (≈ 6) copper ions are located.

One can suppose that all of these copper ions are participating in the formation of nanobridges within a DNA CLCD structure. In this case, each nanobridge, located in each helical turn between the neighboring DNA molecules, contains approximately six copper ions. The evaluation is extremely close to the results of the theoretical calculations [11], showing that one nanobridge can include from four to six copper ions. Consequently, the experimental data and the results of theoretical calculations allow one to accept that a nanobridge $[\ldots\text{-}Cu^{2+}\text{-}DAU\text{-}Cu^{2+}\ldots Cu^{2+}\text{-}DAU\text{-}Cu^{2+}\ldots]$ contains six ions of bivalent copper.

Nanobridges are flat chelate complexes. (See the structures in Figure 4.4, where panel A depicts a nanobridge [view along the DNA molecule axis] and B shows a more detailed structure of nanobridges between DNA molecules that link DNA molecules located both in one layer and in adjacent layers. For the sake of simplicity, the nanobridges are turned 90° relative to their actual orientation.)

The emergence of nanobridges of $[\ldots\text{-}Cu^{2+}\text{-}DAU\text{-}Cu^{2+}\ldots Cu^{2+}\text{-}DAU\text{-}Cu^{2+}\ldots]$ between neighboring DNA molecules can be expected to lead to the formation of a new spatial structure of DNA CLCD. The stability of this structure is determined by the number and properties of nanobridges rather than the properties of the initial polymeric solution. Therefore, as the result of the formation of nanobridges between adjacent DNA molecules, the properties of the initial CLCD particle change, and a structure with new properties evolves [18].

4.4 VISUALIZATION OF RIGID DNA NANOCONSTRUCTIONS LINKED WITH NANOBRIDGES AND MANIPULATIONS WITH THESE NANOCONSTRUCTIONS

The stability of DNA CLCD particles linked with nanobridges does not depend on the properties of the water–polymer solution, but rather is determined by the number and the properties of nanobridges. This means that a new structure would persist even in aqueous salt solution, which substantially facilitates its handling. Taking this circumstance into account, there is an opportunity to visualize the particles of liquid DNA nanostructure after their transformation into a spatially fixed structure.

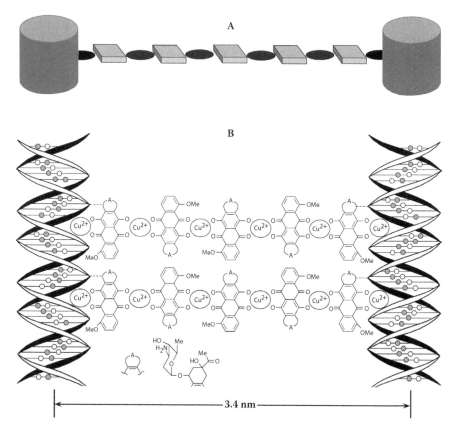

FIGURE 4.4   Structure of nanobridges between two adjacent DNA molecules.

Figure 4.5 shows two-dimensional (2-D) and three-dimensional (3-D) images of the DNA CLCD particles after sequential treatment with DAU and $CuCl_2$ solutions and then immobilized on a nuclear membrane filter. (In Figure 4.5, panels A and B are, respectively, 2-D and 3-D atomic force microscope [AFM] images of the CLCD particles immobilized on a nuclear membrane filter. The dark points are the pores in the filter, and $D \approx 0.2$ μm. Panel C shows the size distribution of DNA CLCD particles cross-linked with nanobridges.)

Examination of the size distribution of particles of nanoconstructions (NaC) obtained by direct measurement of 113 particles showed that the particles have a shape reminiscent of a prolate cylinder, and although their sizes vary from 300 to 700 nm, they have an average diameter of about 500 nm. It should be noted, that this is the value obtained by direct measurement of the dimensions of the DNA CLCD particles. Consequently, the size of the formed CLCD particles with DNA cross-linked with nanobridges coincides with the size of the initial DNA CLCD particles calculated theoretically for the case of solutions with the constant osmotic pressure. This means that the formation of nanobridges between the neighboring DNA molecules results in structures with fixed 3-D spatial configurations. Although

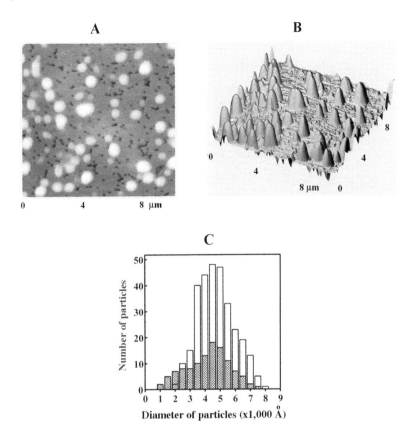

FIGURE 4.5 2-D and 3-D images of CLCD particles with DNA molecules cross-linked with nanobridges and their size distribution.

the number of nanobridges can be insignificant, they can represent the stabilizing factor of the obtained spatial structure. Hence, for the first time, direct data characterizing the macroscopic parameters of particles formed by low-molecular-mass double-stranded DNA molecules were obtained.

The morphology of CLCD particles formed on a synthetic double-stranded polynucleotide poly(I)×poly(C) molecule is almost identical to the DNA CLCD. The average size is about 400 nm, which corresponds to the results of the analysis of the size of dispersion particles formed by DNA molecules with a similar molecular mass. Consequently, the use of DNA CLCD particles allows one to create a new material that is easy to manipulate.

Taking into account the significant differences between the liquid and new structures of DNA CLCD, it would be reasonable (to highlight these differences) to use the term *rigid DNA nanoconstruction* (rigid NaC) to designate the obtained material. The creation of a unique device for AFM in Russia with an ability to measure the exact coordinates of a single particle provides principally new possibilities for investigating the rigid NaC. Now it is possible to cut a rigid NaC.

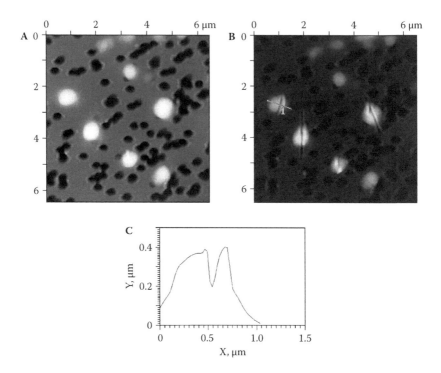

FIGURE 4.6 **(See color insert)** AFM images of rigid nanoconstructions formed by DNA molecules cross-linked with nanobridges and the profile of the cut of the nanoconstruction.

Figure 4.6(A) shows the image of an initial DNA NaC; the image of the same particles after being split (cut) directly on a nuclear membrane filter is shown in Figure 4.6(B); and Figure 4.6(C) gives the cut profile.

The measurements were performed on a scanning probe microscope (SmartSMP) produced by AIST-NY in Zelenograd, Russia. SmartSMP is an automatic probe microscope with a quick 100-μm scanner equipped with capacity movement transducers. To measure the topography of the sample surface, the ACM top mode was used. In this mode, in every spot of the scan, the noncontact cantilever oscillates at a resonance frequency, rises above the surface, and lowers until the amplitude drops to the specified quantity. This way of scanning avoids the generation of parasites on significantly sloped sides of the object that usually evolve when scanning using the traditional semicontact AFM method. A silica noncontact probe with resonance frequency 250 kHz with a carbon whisker on the end was used for scanning. The probe with this kind of configuration is intended for exploring surfaces with sharp or narrow slots and apertures (in this case the apertures on the surface of the filter). To cut particles of NaCs, the probe makes a specified number of movements, pressing the material.

The images of the DNA NaCs along the cut are shown in Figure 4.7. The manipulation on cutting rigid NaCs, realized for the first time, makes it possible to take a new look at the these NaCs and may be quite useful for the creation of matrices

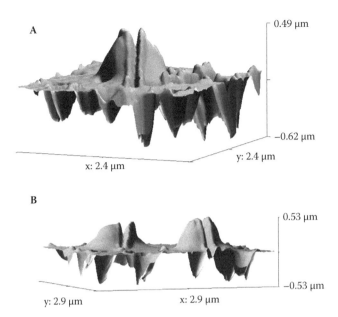

FIGURE 4.7 **(See color insert)** Spatial images of cuts of rigid nanoconstructions formed by DNA molecules cross-linked with nanobridges.

containing a fixed number of rigid NaCs located in a certain mode. Indeed, using the liquid DNA CLCD particle, a new rigid structure was created, as shown in Figure 4.8. This figure shows that in an initial DNA CLCD particle (on left) existing only at a fixed osmotic pressure of a solution (wide arrows) where DNA molecules are ordered, the neighboring DNA molecules were cross-linked with nanobridges without distortion of the total spatial structure of the particle. As a result, it is clear that the formation of the nanobridges has produced a rigid DNA nanoconstruction.

It should be noted that the acquired NaC is unique because it contains not only DNA molecules, but also antibiotic molecules and copper ions, i.e., guest molecules. This structure accommodates a high concentration not only of DNA molecules alone (locally up to 400 mg/mL), but antibiotic molecules (about 200 mg/mL) and copper ions. This means that the considered nanodesign technology automatically solved the problem that remains unsolved in the hybridization technology (chapter 2). Moreover, the formation of NaCs sharply increases the abnormal optical activity in the absorption region of an antibiotic, a circumstance that is very important from a practical point of view, since a high optical activity makes it possible to monitor with high accuracy all changes in the properties of the NaC formed. The diffusion mobility of the DNA molecules in the structure is sharply limited, so the structure does not have many of the properties typical of the initial DNA CLCD particles. The stability of the rigid NaC depends on the properties of the nanobridges and not on the osmotic pressure of the solution. The abnormal optical properties typical of the initial DNA CLCD are retained and even amplified in the rigid NaC. The properties of the rigid NaC provide such possibilities for the practical application of the created structure that cannot be implemented in the case of the formation

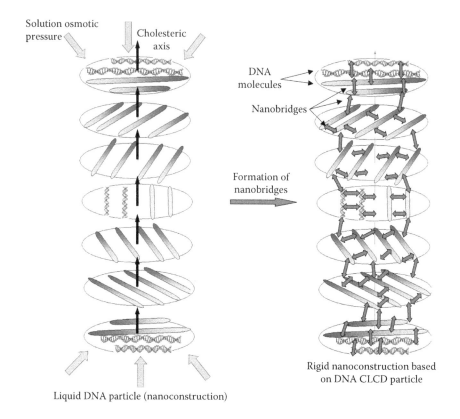

Liquid DNA particle (nanoconstruction)

Rigid nanoconstruction based on DNA CLCD particle

FIGURE 4.8 **(See color insert)** Scheme illustrating the transformation of liquid DNA CLCD particle into rigid DNA nanoconstruction.

of DNA nanostructures by the hybridization technique. This means that the DNA NaC is not only a new type of nanobiomaterial whose properties can be regulated within broad limits but it is also easy to observe any changes of these properties.

4.5 DEPENDENCE OF OPTICAL PROPERTIES OF DNA NANOCONSTRUCTIONS ON TEMPERATURE

If the stability of DNA NaCs depends on the strength and the number of the nanobridges between adjacent DNA molecules, it is obvious that such a structure must have a specific curve of melting temperature.

Figure 4.9 compares the temperature dependencies of the relative values of the band amplitudes in the CD spectrum of the cholesteric LCD of an initial DNA and the resultant NaCs formed under various conditions using this LCD. (In Figure 4.9, curve 1 is for $C_{DAU} = 15.1 \times 10^{-6}$ mol/L and $C_{Cu} = 5.02 \times 10^{-6}$ mol/L; curve 2 is for $C_{DAU} = 27.3 \times 10^{-6}$ mol/L and $C_{Cu} = 9.9 \times 10^{-6}$ mol/L; and curve 3 is for $C_{DAU} = 63.8 \times 10^{-6}$ mol/L and $C_{Cu} = 20.1 \times 10^{-6}$ mol/L. Conditions: $C_{DNA} = 5.5$ µg/mL; $C_{PEG} = 170$ mg/mL; 0.3 mol/L NaCl; 0.002 mol/L phosphate buffer; pH 6.8.) The

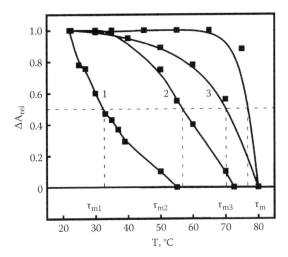

FIGURE 4.9 Dependence of the relative amplitude of the bands in the CD spectra of DNA CLCD (curve without number) and DNA nanoconstructions (curves 1–3) on the temperature.

stability of both cholesteric DNA LCD and NaCs was characterized by the value of the CD-melting temperature (τ_m), conditionally accepting that this is a temperature at which the amplitude of an abnormal band in the CD spectrum decreases by 50%.

The unnumbered curve in Figure 4.9 shows the temperature changes in the abnormal optical activity of the initial DNA CLCD particles without nanobridges (the CD-melting curve). It shows that the DNA CLCD particles are melted in a narrow temperature range ($\approx$5°C) at a τ_m value of about 75°C. The CD-melting of the DNA LCD particles reflects the change in the twist angle of their spatial structure; it is not accompanied by separation of the DNA chains and, due to steric reasons, the process of the CD-melting of the DNA LCD is fully reversible. Curves 1–3 in Figure 4.9 describe the temperature decrease in the abnormal optical activity of DNA NaCs in the absorption region of nanobridge chromophores ($\lambda \approx 510$ nm). The melting of NaCs occurs in the temperature range where the cholesteric structure of the LCD particles still persists. The τ_m values registered at λ 505 nm and at λ 310 nm coincide. Moreover, the CD-melting of NaCs registered by both bands depends on the concentrations of both Cu^{2+} ions and DAU molecules in solution: The higher the concentration of any of these components of the nanobridges (at constant concentration of the other one), the higher the τ_m value of NaCs, though it remains below the τ_m of the initial DNA cholesteric structure [19].

It can also be seen that two thermal structural transitions are typical of DNA NaCs: One of them corresponds to the CD-melting of the nanobridges, and the other corresponds to the CD-melting of the DNA cholesteric structure. The difference in the melting temperatures for DNA cholesteric structures and NaCs based on DNA reflects dissimilar processes. In particular, the melting of the DNA cholesteric structure is explained by the unwinding of the helical cholesteric structure of LCD particles, whereas the melting of the NaCs may be associated with dissociation of the nanobridges into components occurring in the core of intact DNA cholesteric structures upon temperature increase.

Figure 4.9 is consistent with an increase in the number or in the thermodynamic stability of nanobridges formed between the DNA molecules. In addition, a similar character of the dependencies of τ_m on Cu^{2+} and DAU concentration, i.e., upon the concentration of both building elements, suggests that approximately equal amounts of Cu^{2+} ions and DAU molecules play a role in the structure of nanobridges.

The dependence of the τ_m values of NaCs on the number of their integral elements is a typical property of these objects. Assuming that a nanostructure is formed by nanobridges within the NaC, the melting of this structure, as well as melting of other nanoobjects, must be described by the Gibbs–Thomson equation given in chapter 1 (Equation 1.3).

Figure 4.10 illustrates the dependence of τ_m of NACs on the effective length of a nanobridge, i.e., the length per one DNA helical turn. It can be seen that, in accordance with the Gibbs–Thomson equation, the dependence is a straight line.

Figure 4.11 compares the dependencies reflecting the melting of nanoobjects formed by various materials. (In Figure 4.11, curve 1 shows the efficient length of the nanobridges in an NaC; curves 2 and 4 show the respective size of gold and tin nanoparticles; curve 3 shows the length of the ribs of nanometric tetrahedrons from double-stranded DNA fragments; and curve 5 shows the length of an oligo(A)×oligo(U) oligonucleotide chain.) In all cases of nanostructure formation, regardless of the material, the Gibbs–Thomson equation describes the observed effect. The comparison of the given lines verifies the previously asserted conclusion about the universal character of the physical laws describing the properties of nanoparticles (nanostructures), regardless of the material these objects are made of.

Some important properties of rigid DNA NaCs can be enumerated: First, unlike the initial liquid DNA CLCD particles, the properties of rigid NaCs do not depend on the osmotic pressure of the solution. They are determined by the number and the energy of the nanobridges. The presence of the nanobridges cross-linking the

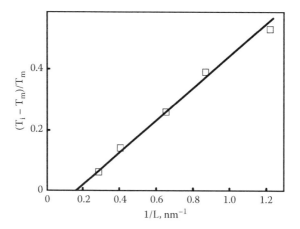

FIGURE 4.10   Dependence of $(T_i - T_m)/T_m$ on $1/L$ for nanoconstructions.

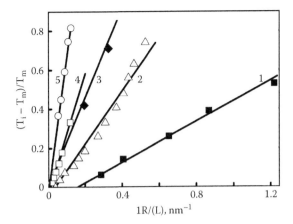

FIGURE 4.11 Dependence of $(T_i - T_m)/T_m$ on $1/R(L)$ for different nanoobjects.

DNA molecules leads to dramatic changes in the properties of the initial DNA CLCD. In particular, the liquid mode of packing of the neighboring DNA molecules disappears, as does the diffusion mobility of the DNA molecules: The structure of the NaC becomes rigid, and the particles possess the features of a solid material. Second, the amplification of the abnormal optical activity of NaCs shows itself by an increase in the amplitude of the bands located in the CD spectrum in the absorption regions of both DNA, and the antibiotic is specific. Third, NACs are described by two thermal structural transitions: One of them corresponds to the CD-melting of the nanobridges, and the other one corresponds to the CD-melting of the DNA cholesteric structure. Finally, NaCs are characterized by a high local concentration of both DNA molecules and an antitumoral antibiotic, DAU.

The structural elements of DNA NaCs are DNA molecules spatially ordered both in one layer and in the neighboring layers as well as the nanobridges connecting them consisting of alternating molecules of an anthracycline antibiotic and copper ions. This unique nanostructure created from nanometric building blocks has no natural analogues.

The unique spatial structure of NaCs—the variety of the building blocks (DNA molecules, copper ions, anthracycline molecules) and the dependence of the NaC properties on a broad number of factors (ionic strength, temperature, concentration of the polymer, and copper and anthracycline concentration)—make it possible to manipulate the properties of the NaC.

Therefore, as the result of the application of the proposed technology, a new rigid biomaterial with tailored properties has been obtained. This technology can be improved by selecting other nanobridge components or using molecules of other nucleic acids, including their complexes with synthetic or natural polymers. It can also be noted that the material produced is called by different names in different works (nanoconstruction, nanomaterial, liquid-crystalline elastomer, rigid cholesteric structure, etc.), and the question of its name is still subject to discussion.

The successful implementation of the proposed approach allows one to assume that there may be other methods to affect the DNA molecules in DNA CLCD particles to induce the transition of the particles to a solid state.

One can add that it is quite clear that destruction of the integrity of nanobridges between DNA molecules under the action of various chemical or biologically relevant compounds will result in disintegration of the spatial structure of rigid DNA NaC. This process should be accompanied by a decrease in the value of the easily detected CD band located in the region of absorption of structural elements of nanobridges. This means that rigid DNA NaC can be used as a microscopic-size multifunctional sensing unit (chip) for biological or chemical needs.

4.6 RIGID NANOCONSTRUCTION FORMED AS A RESULT OF FORMATION OF COMPLEXES OF DNA MOLECULES FIXED IN QUASINEMATIC LAYERS WITH RARE-EARTH METAL CATIONS

Coming back to the quasinematic layer of the DNA molecules in a CLCD particle (Figure 3.13), a number of additional peculiarities of the DNA molecules in a layer can be pointed out. The consideration of these peculiarities would make it possible to develop a second approach to the creation of rigid DNA nanoconstructions. The approach is based on the following facts:

1. Some chemically and biologically active compounds may cause local (nanometric) alterations of the mutual orientation of the adjacent base pairs as they interact with DNA nitrogen base pairs, which leads, on the whole, to the appearance of structural heterogeneities in the secondary structure of DNA.

2. The steric limitations—resulting both from dense packing of the DNA molecules in quasinematic layers and the fixed osmotic pressure of PEG solution—prevent full separation (denaturation) of strands of the DNA molecules ordered in particles of the CLCDs [20]. This means that the action of various compounds on DNA must not only induce the alterations in the secondary structure of these molecules, but also transform into the changes in the interaction mode of neighboring DNA molecules (fragments) in quasinematic layers.

3. Positively charged counterions may efficiently neutralize negative charges of DNA phosphate groups. Under these conditions, two parameters play an important role: the efficiency of the interaction between the counterions and the DNA phosphate groups, carrying the negative charges, and the solubility of [DNA-counterion] complex in water–salt solution. A DNA molecule has a polyphosphate nature, and its solubility is connected to the extent of neutralization of phosphate group negative charges, because a completely neutralized phosphate group of a polyphosphate chain is water incompatible [21, 22]. The extent of neutralization of the phosphate groups depends on the charge value and the origin of the cations interacting with the DNA phosphate groups.

4. At a high osmotic pressure determined by PEG concentration and high ionic strength of the solution, the transition of DNA molecules fixed in the spatial structure of the CLCD particles into an insoluble state must change the mode of the interaction between neighboring DNA molecules. Under these conditions, the particles of the DNA CLCD can acquire a rigid instead of a liquid structure, and the created structure would be able to exist in the absence of the osmotic pressure of a PEG solution.

5. The previously analyzed theory predicts that changes in the interaction between DNA molecules in quasinematic layers (with an average distance [D] between DNA molecules varying from 3.0 to 5.0 nm) must be followed by changes in the parameters of the CLCD helical structure (in particular, the value of the helical twist) and, consequently, by changes in the abnormal optical activity of the initial CLCD.

All of these facts allow one to formulate the basic approach as follows: As a result of the action of a chemical compound (guest) on the DNA molecules fixed in quasinematic layers of CLCD particles, the efficiency of the interaction between DNA molecules (or fragments of these molecules) in neighboring quasinematic layers needs to be increased so that the created construction acquires the features of a rigid (solid) material and could exist in the absence of the osmotic pressure of a polymer-containing solution. The solution of this problem will mean that, basically, another approach makes it possible to create rigid DNA nanoconstructions capable of accumulating guest molecules.

The analysis of Yevdokimov, Salyanov, and Skuridin [23] has shown that chemically or biologically active compounds (guests) needed for the creation of stable constructions of a new type must fulfill certain requirements. First, the interaction between the guests and the DNA must only be followed by nanometric changes in the secondary structure of DNA that do not distort the optical anisotropy of DNA chromophores (nitrogen bases) in a layer, i.e., the interaction of a guest with DNA must not cause changes in the abnormal optical activity of DNA CLCD. Second, the guest molecules must cause total neutralization of DNA phosphate group negative charges, while the [DNA-guest] complex must be poorly soluble (almost insoluble). Third, the accumulation of the specific structural distortions at the increase in concentration of the guest molecules in CLCD particles must induce the spatial approaching of the fragments of DNA molecules both in one layer and in neighboring layers, must amplify the interactions between them and, consequently, must lead to the disappearance of their diffusion mobility. (At the same time, it is possible that nanometric distortions of the DNA molecule structure may lead to the disappearance of the small-range order in the location of DNA molecules in a layer and, consequently, to the disappearance of the small-angle X-ray scattering reflection.) It can be assumed that the combination of these requirements will lead to the transition of DNA molecules in a layer from a liquid to a rigid state and, moreover, to the formation of a stable nanoconstruction of the new type.

Multiple charged cations, such as Al^{3+} or Fe^{3+}, can neutralize the charges of DNA phosphate groups [24–26], but they exist in a solution as a set of hydrated forms, causing uncontrollable changes in the orientation of the DNA base pairs. At the same time, their complexes with phosphate groups have high solubility and are poorly formed at a high ionic strength of the solution used for the DNA phase exclusion.

To reach the low solubility of the DNA molecules, attention is focused on the cations of rare-earth elements (REE). These cations, first, can neutralize the negative charges of the DNA phosphate groups, while the complexes of REE and phosphate are almost insoluble [27–30]. Second, it follows from theoretical evaluations [31] that triple-charged REE cations can cause local (nanometric) changes in the secondary structure of linear double-stranded DNA molecules when interacting with their phosphate groups and base pairs [32]. Finally, some of the cations, i.e., gadolinium cations, have wide practical applications [33, 34].

So at the development of this second approach to creation of rigid DNA nanoconstructions and the accumulation of the guest molecules, various salts of REE were used [35, 36]. Figure 4.12 exemplifies the CD spectra of DNA CLCD in the absence (curve 1) and in the presence of $GdCl_3$ (curves 2–5). (In Figure 4.12, curve 1 shows gadolinium $r_t = 0$; curve 2 $r_t = 1.65$; curve 3 $r_t = 16.5$; curve 4 $r_t = 32.7$; curve 5 $r_t = 95$. Conditions: $C_{DNA} = 10$ μg/mL; $C_{NaCl} = 0.3$ mol/L; $C_{initial}$ $GdCl_3 = 0.1$ mol/L; $C_{PEG} = 170$ mg/mL; $\Delta A = (A_L - A_R) \times 10^{-3}$ optical units; r_t is the ratio of total $GdCl_3$ molar concentration to the molar concentration of the DNA nitrogen bases.)

The initial DNA CLCD is characterized, as usual, by an abnormal band in the CD spectrum with a maximum at $\lambda \approx 260$ nm. (It is worth recalling that the negative sign of the abnormal band in the CD spectrum indicates that the CLCD particles formed by double-stranded [B-form] DNA molecules have a left-handed helical twist.) Figure 4.12 shows that under low $GdCl_3$ concentration, the amplitude of the abnormal band in the CD spectrum of CLCD is reduced nonsignificantly (curves 1–2). However, at a high concentration of $GdCl_3$, a sharp increase in the abnormal band amplitude in the CD spectrum of the [DNA-Gd³⁺] complex takes place, and the maximum is red-shifted (curve 5).

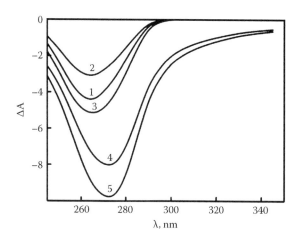

FIGURE 4.12 CD spectra of DNA CLCD formed in PEG-containing water–salt solutions in the absence and presence of $GdCl_3$.

Similar changes in the shape of the CD spectra of DNA CLCD have been observed in cases when the CLCD were added with high concentrations of salts of other REEs (lanthanum, neodymium, samarium, etc.). Moreover, in the case of CLCD formed by double-stranded poly(I)×poly(C) molecules with an abnormal positive band in the CD spectrum, there is an amplification of the band and a red shift of the maximum (data not shown). The treatment of DNA CLCD particles with solutions of trivalent cations of other metals (Fe^{3+}, Al^{3+}, etc.) is followed by a sharp decrease, but not amplification, of the abnormal band in the CD spectrum. Hence, in the case of REEs, there is a specific mechanism that determines the amplification of the abnormal bands in the CD spectra corresponding to the CLCD under a high concentration of the cations.

The observed changes in the shape of the CD spectrum of DNA CLCD (Figure 4.12) are basically similar to the changes of the CD spectra at the formation of the nanobridges between neighboring DNA molecules fixed in quasinematic layers, as discussed previously. The formation of the nanobridges leads to the disappearance of the liquid mode of packing of the DNA molecule in CLCD particles and the formation of a NaC with a rigid spatial structure.

The results obtained by various methods describing the changes in the properties of DNA molecules in gadolinium solutions at different concentrations are compared in Figure 4.13. (In Figure 4.13, the inset compares the dependence of the change in the band amplitude on the $GdCl_3$ r_t value obtained by various methods: curve 1 depicts the CD spectrum of CLCD of [DNA-Gd^{3+}] complex; curve 2 depicts the CD spectrum of linear DNA ($\lambda = 270$ nm); curve 3 shows the dependence of the small-angle X-ray scattering reflection for the phases formed with [DNA-Gd^{3+}] particles. Conditions: $C_{DNA} = 10$ µg/mL; $C_{NaCl} = 0.3$ mol/L; $C_{PEG} = 170$ mg/mL.)

These results show that at the interaction between the gadolinium ions (as well as ions of other REEs) and the DNA molecules fixed in a spatial structure of CLCD particles, two processes at different concentrations of $GdCl_3$ take place. At low $GdCl_3$

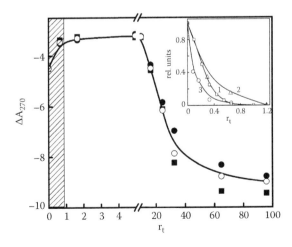

FIGURE 4.13 Changes in the optical properties and X-ray parameters of DNA CLCD at different concentrations of $GdCl_3$.

concentrations, the properties of the initial linear DNA molecules change: First, the amplitude of the *positive band* in the CD spectrum of *the linear double-stranded DNA molecules* decreases, indicating the distortion of the structural parameters of nitrogen bases [32, 37, 38] and, hence, nanoscale modification of parameters of the secondary structure DNA molecules. Second, in the case of densely packed DNA molecules, this process is accompanied by destruction of the parallel orientation of these molecules and, hence, by the disappearance of the X-ray maxima in the scattering curves for the phases obtained by low-speed sedimentation of the CLCD particles of [DNA-Gd^{3+}] complexes [37].

To analyze the second reason more carefully, it is necessary to take into account the following: As shown previously, when the rare earth cations are bound to linear double-stranded DNA molecules, the shape of the CD spectrum of these molecules changes sharply at $r_t \approx 0.5$. This change demonstrates the loss of the regular homogeneous character typical of the secondary structure of DNA. Such alteration can be associated with a local (nanoscale) conformational transition, for instance, of B-Z type. This can be considered as an argument that the binding of gadolinium ions to DNA molecules packed in particles of the CLCD at low concentrations leads to a nanoscale alteration in homogeneity of the secondary structure of the DNA. Such modified nucleic acid molecules are separated into alternating fragments differing in conformations (e.g., -B-Z-B-B-Z-Z-etc. for DNA). The junctions between B-DNA and Z-DNA fragments also contain extruded bases [37], providing the sites with modified nanosize properties.

The existence of modified DNA, containing the B-form fragments, is confirmed by the fact that planar antibiotics (daunorubicin, mitoxantrone) can be incorporated between the base pairs of the modified DNA in the particles of the CLCD of [DNA-Gd^{3+}] complexes. If the fragments of the neighboring molecules of the [DNA-Gd^{3+}] complexes, or even all of the molecules packed in particles of the CLCD, acquire heterogeneous secondary structure, the translational order of these fragments (molecules) is broken, resulting in disappearance of the small-angle reflection on X-ray patterns, as was observed in our experiment (see Figure 4.13). Indeed, the general characteristics of the peaks—such as peak position s_{max} on the scattering patterns, repeating distances of the periodical motifs in the crystalline regions, $d = 2\pi/s_{max}$—mean that long-range order dimension L (the size of crystallites) and the degree of disorder in the system Δ/d do not vary, demonstrating that the structure of parallel packed DNA molecules is constant. However, consecutive reduction of the areas under Bragg peaks after the addition of gadolinium to the DNA CLCD reflects a decrease in the number of ordered quasi-crystalline regions in the whole system. At the same time, the central scattering is intensified, which reflects the formation of a new type of nanoscale structural heterogeneity, which most likely comprises conglomerates of [DNA-Gd^{3+}] complexes without parallel ordering. Thus, X-ray data show that the treatment of the DNA CLCD by gadolinium destroys the typical parallel orientation in the dense packing of DNA molecules and leads to the formation of disordered regions.

The state of the DNA secondary structure under conditions of high concentration of gadolinium cations was checked by application of an "external chromophore" approach based on theoretical consideration of the peculiarities of the CLCD CD

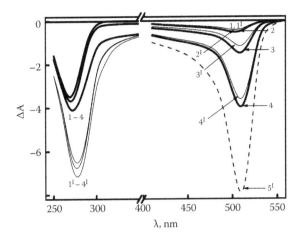

FIGURE 4.14 Comparison of the CD spectra of the DNA CLCD treated with SYBR Green (curves 1–4) to the CD spectra of CLCD formed initially by [DNA-Gd³⁺] complexes and then treated with SYBR Green (curves 1¹–4¹).

spectra [39, 40]. As an external chromophore, we have used again "SYBR Green," an intercalating fluorescent dye [41].

Figure 4.14 demonstrates the obtained results. One can see that in agreement with theoretical predictions [39] the treatment of CLCD particles formed by [DNA-Gd³⁺] complexes with SYBR Green is accompanied by an appearance of an additional abnormal negative band in the CD spectrum. (In Figure 4.14, the dotted curve 5′ depicts the expected CD spectrum of the CLCD formed by [DNA-Gd³⁺] complexes and treated with SYBR Green under the condition of homogeneity in the DNA secondary structure. SYBR Green r_t in the cases of curves 1 and 1¹ = 0; in the cases of 2 and 2¹ = 0.033; in the cases of 3 and 3¹ = 0.2; in the cases of 4 and 4¹ = 0.66; and in the case of 5¹ = 0.66. Conditions: C_{DNA} = 10 μg/mL; C_{PEG} = 170 mg/mL; C_{NaCl} = 0.3 mol/L; C_{GdCl3} = 0.003 mol/L; $\Delta A = (A_L - A_R) \times 10^{-3}$ optical units. SYBR Green r_t - is the ratio of total SYBR Green molar concentration to the molar concentration of the DNA nitrogen bases.)

Figure 4.14 shows that there are two bands in the CD spectrum of [DNA-Gd³⁺] complexes. One occurs in the absorption region of the DNA nitrogen bases, and the other lies in the absorption region of SYBR Green chromophores (≈510 nm). Under binding of SYBR Green with double-stranded DNA molecules in the CLCD particles formed by [DNA-Gd³⁺] complexes, both bands have negative signs despite the SYBR Green concentration. The identical signs of two bands in the CD spectra unequivocally mean that SYBR Green molecules are fixed in quasinematic DNA layers. The amplitude of the band in the CD spectrum in the region of SYBR Green absorption grows with an increase in the number of its molecules bound to DNA, although the amplitude of the band in the region of DNA absorption remains practically constant. These CD spectra mean that the orientation of SYBR Green molecules coincides with the orientation of the nitrogen base about the DNA axis and that SYBR Green molecules intercalate into DNA so that the angle between SYBR Green molecules

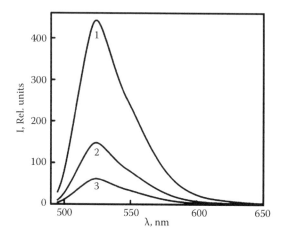

FIGURE 4.15 Fluorescence spectra of the initial, linear DNA and CLCD formed by DNA as well as the spectrum of CLCD of [DNA-Gd^{3+}] complexes treated by SYBR Green (curves 1, 2, and 3, respectively).

and the long axes of the DNA is ≈90°. However, the experimentally measured amplitudes (compare, for instance, curve 4^1 to theoretically calculated curve 5^1) are two times smaller than the expected ones (if one can take into account the correlation between the amplitudes of abnormal bands in the DNA and the absorption regions of the external chromophores [39]). The detected difference shows that, indeed, at high gadolinium concentration, the breaking of the DNA secondary structure in the CLCD particles of [DNA-Gd^{3+}] complexes takes place. The measurements of the SYBR Green fluorescence in the content of CLCD particles formed by [DNA-Gd^{3+}] complexes support this point of view.

Figure 4.15 demonstrates that the fluorescence intensity of SYBR Green is decreased in the case of both DNA CLCD and [DNA-Gd^{3+}] CLCD. (In Figure 4.15 C_{DNA} = 10 μg/mL; C_{NaCl} = 0.3 mol/L; C_{PEG} = 170 mg/mL; $C_{SYBR\ Green}$ = 9.73 × 10^{-6} mol/L; C_{GdCl3} = 0.003 mol/L.) The most important facts consist in the following: (a) condensation of DNA molecules and formation of the DNA CLCD is accompanied by a drop in the intensity of fluorescence of SYBR Green, and (b) there is difference in the fluorescence of SYBR Green intercalated between nitrogen bases pairs in the content of DNA molecules ordered in initial CLCD and in the content of CLCD treated with gadolinium. Since the solvent used for measurements of curves 2 and 3 was not changed and because SYBR Green molecules bind with regular B-form of DNA, the difference between curves 2 and 3 confirms once more the statement that neighboring DNA molecules, packed in particles of the CLCD [DNA-Gd^{3+}] complexes, acquire an inhomogeneous secondary structure.

Concerning the received results, two related questions evolve. First, why under the conditions of GdCl$_3$ r_t < 1, does the amplitude of the abnormal band in the CD spectrum of the CLCD of [DNA-Gd^{3+}] complex remain almost constant? Second, why does the amplitude of the abnormal band increase at GdCl$_3$ r_t > 20?

In the theories describing the abnormal optical activity of low-molecular-mass cholesteric structures [42–46], the existence of an ordered location and orientation of chromophores in quasinematic layers and helical ordering of neighboring quasinematic layers formed by these chromophores is enough for an appearance of abnormal optical activity. As a rule, the theories do not consider the possible correlation between the crystallographic packing of neighboring molecules and the orientation order of their chromophores in the cholesteric structure. Apparently, the absence of such correlation is caused by the fact that, in the case of low-molecular-mass compounds, there is a definite connection between the direction of the long axis of the molecule and the orientation of the chromophore. In the case of DNA molecules with a helical structure, the correlation between the orientation of chromophores (nitrogen bases) in respect to the long axes of the molecules in a quasinematic layer and the orientation order of nitrogen bases in the CLCD particle structure can be most complicated. However, just as in the case of low-molecular-mass compounds, in the case of DNA cholesteric structures, a high density of chromophore (nitrogen base) packing is required for the abnormal optical activity to appear [44]. If the packing density and the orientation order of the chromophores (nitrogen bases) of DNA in quasinematic layers of the CLCD particles is not distorted, the amplitude of the abnormal band in the CD spectrum remains constant, which has been observed experimentally (Figure 4.13).

Based on the data obtained using various physical methods, when the negative charges of DNA phosphate groups are neutralized by positively charged gadolinium ions under conditions of high gadolinium concentration [35], a different process takes place. First, the gadolinium ions neutralize the negative charges of DNA phosphate groups and make the CLCD particles insoluble. Indeed, under conditions of an excess of gadolinium ions (Figure 4.13), these ions displace the sodium ions and bond to the DNA phosphate groups. When the Gd^{3+} ions are bonded to the polyphosphates, a poorly soluble gadolinium polyphosphate is formed (solubility constant is below $\approx 10^{-(13-14)}$ mol/L) [27–30, 47]. As molecules of double-stranded DNA are polyphosphate by nature, these molecules become poorly soluble both in water–salt and in PEG-containing solutions at a high concentration of gadolinium. This means that a stable spatial structure can be formed under such conditions, and the presence of PEG as a factor stabilizing the structure is not required.

The control experiments have shown that a tenfold dilution of a PEG-containing (C_{PEG} = 170 mg/mL) solution, where CLCD particles are formed from [DNA-Gd^{3+}] complexes, does not cause any changes in the abnormal optical activity of CLCD (taking into account the decrease in the CLCD particle concentration as a result of the dilution). Moreover, the gadolinium ions, which neutralize the negative charges of DNA phosphate groups, create an excessive positive surface charge on the CLCD particles, and aggregation of these particles in the solution becomes impossible. The DNA CLCD particles can sediment in the solution after the decrease in their solubility, which leads to the formation of a pellet. This process results in a decrease in the amplitude of the abnormal band in the CD spectrum. However, an intensive shaking of the solution containing the formed pellet leads to a complete restoration of the initial abnormal optical activity. This means that, as a result of the interaction of gadolinium ions with the DNA molecules in the content of the CLCD particles, these particles lose their ability to coalesce.

Therefore, the DNA CLCD particles, whose phosphate groups are neutralized by gadolinium ions, become poorly soluble, and they can exist in the absence of the high osmotic pressure of a PEG-containing solution (induced by high PEG concentration). This means that the osmotic pressure of a water–salt (or water) solution is sufficient to maintain the helical spatial packing of the $[DNA-Gd^{3+}]$ complexes in CLCD particles, which substantially facilitates their handling.

4.7   VISUALIZATION OF RIGID PARTICLES OF CHOLESTERIC LIQUID-CRYSTALLINE DISPERSION OF A [DNA-GD³⁺] COMPLEX

As noted previously, the particles of the CLCD of the initial DNA cannot exist in the absence of high osmotic pressure of the solvent. Therefore, their fixation is on a nuclear membrane filter, and visualization is impossible under these conditions. However, if poorly soluble CLCD particles, consisting of the $[DNA-Gd^{3+}]$ complexes are formed, then the immobilization of these particles on the surface of the nuclear membrane filter becomes possible, and the size and shape of these particles can be investigated.

Figure 4.16 displays the images of CLCD particles formed by $[DNA-Gd^{3+}]$ complexes immobilized on the surface of a nuclear membrane filter. (In Figure 4.16, C_{DNA} = 1.07 µg/mL; C_{NaCl} = 0.03 mol/L; C_{PEG} = 17 mg/mL; C_{GdCl3} = 0.23 × 10^{-3} mol/L. Two sites of the filter are shown.) It is evident that these particles exist as independent individual objects, which are easy to visualize. The existence of the independent particles testifies in favor of the assumption of the appearance of a noncompensated surface charge on the CLCD particles of $[DNA-Gd^{3+}]$ complexes, preventing them from the coalescence.

Figure 4.17 demonstrates the size distribution of these particles as well as the pores in the filter. (In Figure 4.17, C_{DNA} = 1.07 µg/mL; C_{NaCl} = 0.03 mol/L; C_{PEG} = 17 mg/mL; C_{GdCl3} = 0.23 × 10^{-3} mol/L.)

The presence of single particles in Figure 4.16 demonstrates that, upon treatment of particles of DNA CLCD by $GdCl_3$, the liquid character of the DNA packing in these particles disappears, and the particles have a rigid spatial structure. Therefore, particles of the CLCD of the DNAs whose phosphate groups are neutralized by gadolinium ions become poorly soluble and can exist in the absence of osmotic pressure of the PEG-containing solution, and the osmotic pressure of the water–salt PEG-containing solution is not required to support the spatial structure of CLCD particles formed by $[DNA-Gd^{3+}]$ complexes.

The mean particle size is 450–500 nm, i.e., the mean diameter of the DNA CLCD particles after gadolinium treatment coincides with the mean diameter of initial DNA CLCD particles [48]. The particles have the shape of spherocylinders, and the diameter of the particles is close to their height. The obtained result is important, because it allows one to suggest that the average packing density of the DNA molecules in particles of the CLCD of the $[DNA-Gd^{3+}]$ complex is quite close to the

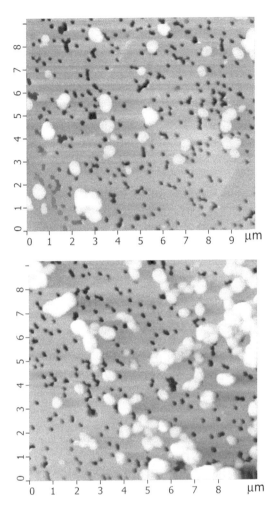

FIGURE 4.16 AFM images of the DNA CLCD particles formed [DNA-Gd³⁺] complexes immobilized onto the surface of nuclear membrane filter. (The dark spots are pores in the filter.)

packing density of the DNA molecules in particles of the CLCD formed from initial DNA molecules. In this case, the mean concentration of chromophores (nitrogen bases) of the DNA in particles of the CLCD of the [DNA-Gd³⁺] complex must also remain not only high, but be sufficient to hold the abnormal optical activity of these particles [49].

Hence, the treatment of the DNA CLCD by GdCl₃ is accompanied not only by neutralization of phosphate groups of the DNA molecules by Gd³⁺ ions, but by a significant attraction between the neighboring DNA molecules. Disappearance of the fluidity of the DNA CLCD particles proves an attractive interaction between the charged DNA molecules arising from interlocking Gd³⁺ ions, sometimes called *counterion cross-links.*

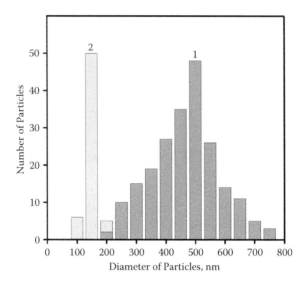

FIGURE 4.17 Size distribution of CLCD particles of [DNA-Gd^{3+}] complexes (1) and the pores in a nuclear membrane filter (2).

The existence of stable CLCD particles of [DNA-Gd^{3+}] complexes made it possible to visualize them with another method. Figure 4.18 shows an image of the CLCD particles of [DNA-Gd^{3+}] complexes treated with SYBR Green. (In Figure 4.18, C_{DNA} = 10 µg/mL; C_{NaCl} = 0.3 mol/L; C_{PEG} = 170 mg/mL; C_{GdCl3} = 2.92 × 10^{-3} mol/L; SYBR Green C_t = 9.8 × 10^{-6} mol/L.)

It can be seen that, despite the formation of the complex between DNA molecules and gadolinium ions, the particles are characterized by a significant fluorescence, which allows one to detect them even in the conditions of a PEG water–salt solution with a confocal fluorescent microscope. Again, [DNA-Gd^{3+}] CLCD particles exist as individual independent objects, i.e., the coalescence of these particles is difficult. The average size of the CLCD particles in PEG water–salt solution evaluated by fluorescent microscopy data almost coincides with the sizes of the particles evaluated by theoretical methods. This size is also very close to the size of the same particles immobilized on a nuclear membrane filter (Figure 4.16).

The combination of the data on the visualization of individual CLCD particles of [DNA-Gd^{3+}] complexes obtained by different methods indicates that the liquid character of packing of DNA molecules in these particles is indeed eliminated, and the particles acquire a rigid spatial structure. Such a structure demonstrates not only the decrease in the solubility of DNA molecules, but also the presence of strong interaction between the fragments of neighboring DNA molecules, because gadolinium ions can be nonuniformly distributed.

The results of low-temperature magnetometric studies showed that at r_t > 20, one gadolinium cation is bounded approximately to one DNA phosphate group, i.e., each phosphate group of DNA molecules, carrying one "effective" negative charge, is neutralized by the Gd^{3+} ion, which carries three positive charges. Hence, at a

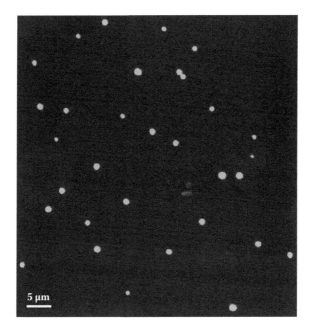

FIGURE 4.18 (**See color insert**) Image of CLCD particles of [DNA-Gd³⁺] complexes in water–salt PEG-containing solution treated with SYBR Green. (Image was taken by Leica TCS SP5 confocal microscope.)

high extent of gadolinium binding to the DNA, not only are the negative charges of phosphate groups neutralized by the Gd cations, but the altered surface charge distribution also makes an additional contribution to the chiral interaction between neighboring molecules of [DNA-Gd³⁺] complexes in the particles.

Gadolinium ions, when present at high concentration, are capable of overcompensating the DNA charges, and a DNA charge inversion can take place, inducing a change in the spatial structure of the CLCD. Indeed, the separation of chains of the DNA molecules in the content of particles of the CLCD is impossible due to steric reasons [20, 50]. The synergic effect of the nanoscale alterations of the secondary structure and the charge inversion of the DNA molecule change the mode of spatial packing of these molecules in the particles of CLCD.

Under conditions of dense packing, the electrostatic forces between DNA molecules, which are intimately connected with a helical mode of charge distribution, are large enough to overcome the intrinsic forces that stabilize the initial cholesteric structure specific to B-form structure [51–54]. This can alter the helical twisting of the cholesteric structure of CLCD particles of [DNA-Gd³⁺] complexes, i.e., facilitate the change in their cholesteric pitch value, P.

To confirm this statement, theoretical evaluations based on the theory of Belyakov et al. [39] were performed. This theory makes it possible to describe the optical properties of particles of DNA CLCD treated with GdCl₃, in particular the correlation between an abnormal optical activity of the DNA CLCD and the P value of its cholesteric structure. Assuming that the diameter of the particles, D, is about 400 nm

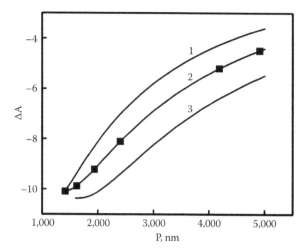

FIGURE 4.19 Theoretical dependencies of the maximal amplitude, ΔA, of the band in the CD spectra of the DNA CLCD treated with GdCl$_3$ upon the pitch, P, value. (The black points correspond to the experimentally measured ΔA values.)

in the case of the DNA CLCD, the P value of the initial cholesteric structure (in the absence of gadolinium ions) is about 3 µm, and the initial ΔA value in the CD spectrum of DNA CLCD is $4,000 \times 10^{-6}$ optical units, which corresponds to the theoretical curves in ($\Delta A - P$) coordinates that were previously calculated. Figure 4.19 shows some of the calculated dependences. (In Figure 4.19, curve 1 depicts the diameter of the particles for $D = 350$ nm; curve 2 shows diameter for $D = 450$ nm; curve 3 shows the diameter for $D = 550$ nm. Conditions: $C_{DNA} = 10$ µg/mL; $C_{NaCl} = 0.3$ mol/L; $C_{PEG} = 170$ mg/mL; $\Delta A = (A_L - A_R) \times 10^{-3}$ optical units.)

These curves clearly show that the decrease in P value is accompanied by an increase in ΔA value, i.e., the smaller the pitch of a cholesteric structure, the greater is the twist, and the more intense is the amplitude of an abnormal band in the CD spectrum. The experimental meanings of ΔA, i.e., the observed amplitudes of the abnormal band in the CD spectra of the CLCD formed by DNA and treated with GdCl$_3$, correspond fairly well to the theoretical curve for the particles with a diameter of 450 nm (Figure 4.19). Comparison of the experimental results and the theoretical calculation confirms the supposition about the correlation between the decrease in P value and the increase of an abnormal optical activity of the CLCD particles treated by the gadolinium salt.

Then, taking into account the known concentration of gadolinium, a correlation between the P values and gadolinium concentration in the solutions was estimated, as seen in Figure 4.20. (In Figure 4.20, curve 1 is for $D = 350$ nm; curve 2 is for $D = 450$ nm; curve 3 is for $D = 550$ nm. Conditions: $C_{DNA} = 10$ µg/mL; $C_{NaCl} = 0.3$ mol/L; $C_{PEG} = 170$ mg/mL.) Figure 4.20 shows that the higher the concentration of gadolinium in the solution, the smaller the P value, i.e., the more twisted the helical structure formed by molecules of [DNA-Gd^{3+}] complexes in CLCD particles. This means that the P effect exists in the system under study, i.e., the increase in the

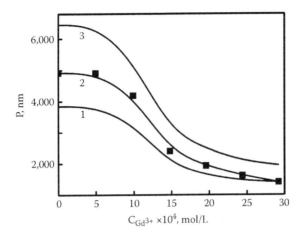

FIGURE 4.20 Theoretical dependencies of the P value of the cholesteric structure for the DNA CLCD particles upon the $GdCl_3$ concentration.

abnormal band in the CD spectrum of DNA CLCD is connected to the change in the twist angle of neighboring DNA layers in the structure of the CLCD particle. Hence, at a high extent of bonding of gadolinium to DNA, not only are the negative charges of phosphate groups neutralized by gadolinium cations, but the changed distribution of surface charges also contributes to the chiral interaction between neighboring molecules in the [DNA-Gd^{3+}] complexes [53, 54].

It is important to emphasize that the problem of the correlation between the local heterogeneity of the structure of nucleic acid molecules and the mode of their ordering in densely packed liquid-crystalline structures remains a subject of active theoretical investigation [39, 53–58]. A change in the twisting of the cholesteric structure can occur in the form of a phase transition. It is worth noting that the possibility of a phase transition between two cholesteric forms was recently analyzed by Golo, Katz, and Kikot [59]. However, such a transition in our case can be quite diffuse (Figure 4.20) due to the smallness of the system elements.

Thus, the combination of two different effects determines the unique optical properties of the CLCD of the complexes of the DNA with gadolinium. The first effect is the formation of nanoscale heterogeneities in the DNA secondary structure at the interaction with gadolinium. The second effect is the enhancement of the interaction between the modified DNA molecules. The first effect is responsible for the "nematization" of the properties of the DNA molecules in quasinematic layers in the structure of the particles of CLCD formed by DNA-gadolinium complexes, i.e., for the disappearance of the small-angle X-ray scattering (SAXS). The second effect is responsible for the large twisting of the cholesteric spatial structure of the CLCD particles, i.e., the amplification of the negative band in the CD spectrum. Similar unique properties are characteristic of the CLCD particles of DNA treated with other REE cations (lanthanum, neodymium, holmium, and ytterbium) as well as for poly(I)×poly(C) CLCD particles added with gadolinium salts.

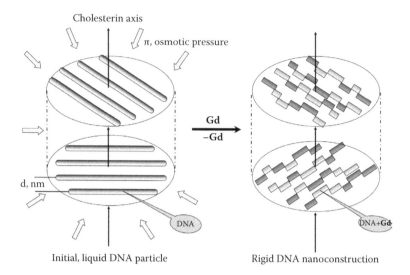

Cholesterin axis

π, osmotic pressure

Gd

−Gd

d, nm

DNA

DNA+Gd

Initial, liquid DNA particle   Rigid DNA nanoconstruction

FIGURE 4.21 **(See color insert)** Scheme of transition of the structure of CLCD particle from a liquid to a rigid state induced by a high concentration of gadolinium ions. (The left panel represents cholesteric-1; the right panel represents cholesteric-2.)

The experimental results suggest a mechanism for formation of a new type of rigid DNA nanoconstruction (Figure 4.21). These results show that REE cations interacting with nucleic acid molecules fixed in the structure of particles of the CLCD induce nanoscale alterations in the secondary structure of these molecules. The accumulation of these alterations is accompanied by a change in the DNA–DNA interactions, resulting in a spontaneous change in the helical twisting of the cholesteric structure of the CLCD. This means that the initial, liquid, cholesteric-1 (Figure 4.21), determined by the combination of properties of the initial nucleic acid molecules (in particular, the conformation of the nucleic acid, charge of phosphate groups, the mode of charge distribution in the DNA helical grooves, etc.) is spontaneously transformed to a more strongly twisted cholesteric-2 (Figure 4.21), which depends on a new set of properties for the modified nucleic acid. One can see that in the case of cholesteric-2, the spatial ordering of neighboring DNA molecules in quasinematic layers is practically absent. Moreover, under these conditions, the twist angle between neighboring DNA quasinematic layers is increased (the P value is decreased, as seen in the right-hand structure). Change in the DNA cholesteric pitch implies a reorganization of the whole spatial structure of the particles of the CLCD. Therefore, a DNA machine, i.e., a particle of the CLCD, consisting of 10^4 spatially ordered nucleic acid molecules, makes work. This means that there is a possibility for transforming the energy of the interaction between gadolinium ions and nucleic acid molecules into the work of the microscopic machine consisting of nucleic acid molecules fixed in the spatial structure of particles of the CLCD, which is easily detected by the change in the abnormal optical activity.

However, the strong interaction between the fragments of neighboring DNA molecules can result not only in stabilization, but a decrease in the solubility of the whole structure. The rigid structure appears, and it can exist in the absence of osmotic pressure of the solvent. Hence, a transition between two spatial structures of CLCD particles occurs. This transition leads to the appearance of a rigid CLCD structure instead of the initial liquid DNA CLCD structure (Figure 4.21).

This new structure of DNA CLCD particles has a unique set of properties:

1. According to the results of magnetometric and neutron-activation analysis [35, 36], the local concentration of gadolinium ions in the received rigid particles is quite high.

2. The heterogeneous nature of the nitrogen base pairs in the DNA molecules suggests that the interaction between REE cations and DNA molecules is followed by nanometric changes in the conformation (of B $\rightarrow$ Z type) only in certain fragments of the molecule [38]. Under these conditions, the homogeneous character of the secondary structure of DNA molecules fixed in the quasinematic layers is distorted, which shows itself through the decrease (up to complete disappearance) of the maximum on the curves of the small-angle X-ray scattering curves as a result of treatment of DNA CLCD particles with gadolinium salts.

3. The interaction between DNA molecules (fragments) with different conformations in quasinematic layers is enhanced.

4. Under gadolinium saturation conditions, the DNA molecules lose their solubility. Due to the fixed osmotic pressure of the PEG solution, they cannot abandon the CLCD particles, and the loss of solubility of individual DNA molecules in the quasinematic layers is followed by a loss of solubility of the whole particle. This induces the transition of the liquid structure of the particles to a rigid state.

5. Due to the high positive charge of gadolinium ions neutralizing the negative charges of DNA phosphate groups, the received rigid particles carry a noncompensated positive charge that prevents them from coalescing.

6. The average size of the CLCD particles formed by [DNA-Gd^{3+}] complexes that reaches 400–500 nm almost coincides with the size of the initial DNA CLCD particles. This result indicates that the average density of packing of the DNA molecules in DNA-Gd^{3+} complex CLCD particles is close to the packing density of DNA molecules in CLCD particles formed from the initial molecules.

7. The retained high average density of the chromophores (nitrogen bases) of DNA [44] in CLCD particles of [DNA-Gd^{3+}] complexes is a condition for the abnormal optical activity of these to be retained.

8. Considering the previous statement, the change in the type of interaction between DNA molecules in quasinematic layers is followed by a decrease in the pitch of the helical twist typical of the spatial structure of DNA CLCD particles [37] and determines a sharp increase of the abnormal optical activity that shows itself through the negative band in the CD spectrum in the absorption region of DNA chromophores as the particles are transformed to the rigid state.

Taking into account these specific properties, it would be reasonable to call the rigid structure based on [DNA-Gd^{3+}] complexes a "rigid DNA NaC." In this case, the NaC spatial properties are not strongly different from the peculiarities of NaCs formed by linking the neighboring DNA molecules with nanobridges. The unique features describing the new DNA NaC mean that the second approach has created a structure containing a high concentration of guest molecules (in the considered case, gadolinium ions). Consequently, a DNA NaC containing a high concentration of REEs has been created, and it has the potential for practical applications. Preliminary experiments have shown the high efficiency of the application of CLCD particles of [DNA-Gd^{3+}] complexes for the destruction of living cells under the effect of the secondary radiation of gadolinium induced by a thermal neutron flow [36, 60].

4.8   AU NANOPARTICLES CAN INDUCE FORMATION OF RIGID DNA NANOCONSTRUCTION

The unique properties of Au nanoparticles were enumerated in Chapter 1. The spherical gold nanoparticles (sometimes called gold nanodots or gold nanoclusters) attract the attention of researchers because they are chemically stable and show unique optical properties. The plasmon resonance band of the gold nanodots has its peak at $\lambda \approx 550$ nm, depending on both the size of the particles and the dielectric constant of the medium [61].

Although massive gold is a safe material, the issue of biocompatibility of nanometric Au particles and how they affect the environment is a matter of interest. The answer to this question is important if a large amount of particles are supposed to be produced for an in vivo application [62, 63].

Chapter 3 discussed the properties of DNA CLCD, which can be considered as a simple model for organizing DNA molecules in the content of the chromosomes of primitive biological objects. The properties of this model structure are nicely changed both at the distortion of the DNA molecule secondary structure and at the modification of the structure of quasinematic layers in the CLCD particle.

Taking into account the real size of nanoparticles, for instance, Au nanoparticles are comparable with the distance between DNA molecules closely packed in quasinematic layers and with structural parameters typical of the secondary structure of these molecules. The interaction mode of these nanoparticles with DNA CLCD can be very complex. One can suppose that there are, at a minimum, two main mechanisms of fixation of Au nanoparticles in the DNA CLCDs.

1. Au nanoparticles can occupy a free space available in the DNA secondary structure and can interact with nitrogen bases of DNA molecules; this will cause the distortion of their secondary structure.
2. Au nanoparticles can occupy a free space available between DNA molecules, inducing cross-links; this will lead to the distortion of the structure of quasinematic layers. Based on the results presented in chapter 2, we can state that in the case of any of these mechanisms, the changes in the CD spectra of the DNA CLCD particles are easily predicted.

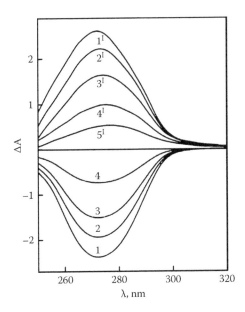

FIGURE 4.22 CD spectra of double-stranded DNA CLCD (curves 1–4) and double-stranded poly(I) × poly(C) CLCD (curves 1^I–5^I) treated with Au nanoparticles (2 nm).

The CD spectra of DNA and poly(I)×poly(C) CLCDs treated with Au nanoparticles (diameter ≈ 2 nm) are compared in Figure 4.22. (In Figure 4.22, curves 1 and 1^I are for $C_{nano-Au}$ = 0; curves 2 and 2^I are for $C_{nano-Au}$ = 0.07 × 10^{14} and 0.16 × 10^{14} particle/mL, respectively; curves 3 and 3^I are for $C_{nano-Au}$ = 0.26 × 10^{14} and 0.33 × 10^{14} particle/mL, respectively; curves 4 and 4^I are for $C_{nano-Au}$ = 0.82 × 10^{14} and 0.66 × 10^{14} particle/mL, respectively; curve 5^I is for $C_{nano-Au}$ = 0.82 × 10^{14} particle/mL. Conditions: C_{DNA} = 9 µg/mL; C_{PEG} = 150 mg/mL; $C_{poly(I)×poly(C)}$ = 9 µg/mL, C_{PEG} = 170 mg/ml; 0.27 mol/L NaCl + 1.78 × 10^{-3} mol/L phosphate buffer; $\Delta A = (A_L - A_R)$ × 10^{-3} optical units. The CD spectra of the DNA CLCD particles were taken 3 h after the addition of gold nanoparticles.)

First of all, one can see that the formation of the CLCDs (compare curves 1 and 1^I) is accompanied by an appearance of abnormal bands in the CD spectra located in the region of absorption of nitrogen bases. The negative sign of the band in the CD spectrum indicates a left-handed cholesteric twist of the right-handed DNA molecules (B-form, curve 1), whereas the positive sign corresponds the right-handed cholesteric twist of the right-handed poly(I)×poly(C) molecules (A-form, curve 1^I) in the formed particles.

Figure 4.22 shows that in both cases, the amplitudes of abnormal bands in the CD spectra drop. The higher the concentration of Au nanoparticles in solution, the greater is the decrease in the abnormal band in the CD spectra of CLCDs of both types of nucleic acids.

There is some question about the probable reasons of the observed optical effect. Based on the data given here, it possible to state at least two reasons that can be used as a basis for the explanation of the observed effect.

1. DNA phosphate groups perform the function of polydentate chelate-forming ligands toward Au nanoparticles, and they also direct the particles into the major grooves of B-form DNA molecules. (B-form DNA has a broad groove with 0.9 nm depth and 1.43 nm width, quite close to the size of a gold cluster with an average diameter of approximately 2 nm.) Calculations have shown that the energy of stabilization of an Au_{55}-nanoparticle (Au_{55}-cluster) in a major groove of B-form DNA is 286 kcal/mol.

2. If Au nanoparticles with a specified size (in this case, approximately 2 nm) can somehow be fixed in the wide groove of neighboring DNA molecules, at a certain concentration of such particles in DNA molecules and, consequently, in a quasinematic layer, then the type of the interaction between adjacent DNA molecules will change. In these conditions, modification of the quasinematic layer may take place, which will lead to a distortion of the cholesteric structure of CLCD particles (the helical twist of the adjacent quasinematic layers will decrease), with a corresponding decrease in the abnormal optical activity of CLCD.

It is obvious that the maximal efficiency of fixation of Au nanoparticles in the wide grooves of DNA molecules will be observed when the size of the particles is below or approximately equal to the size of this groove. Indeed, with an increase in the size of the Au nanoparticles, the effect of the decrease in the abnormal optical activity of CLCD particles decreases, the result of a reduction in the efficiency of the fixation of gold nanoparticles in the wide groove of DNA.

At a particle size of 15 nm, the observed optical effect is almost absent because the penetration of particles of this size into a quasinematic layer is difficult at room temperature and, besides, for steric reasons they cannot be fixed in the wide groove of a DNA molecule. Hence, the structure of a quasinematic layer can efficiently respond to the excitation induced by Au nanoparticles of a specific size.

Finally, we cannot exclude the possibility that another effect following the binding of Au nanoparticles with DNA molecules may present itself. The matter is that Au nanoparticles interacting with negatively charged oxygen atoms in the phosphate groups can cause a local alteration of the B-form DNA structure [64, 65]. At the same time, it is quite probable that the nitrogen-containing groups of purine and pyrimidine bases located in the wide groove of B-form DNA can take part in the interaction (nonspecific binding) with Au nanoparticles [66, 67], which may lead to the distortion of the electron structure and the conformation homogeneity of nitrogen bases (chromophores) along the DNA molecule chain [61]. Under these conditions, the amplitude of the abnormal band in the CD spectrum of DNA CLCD may also be reduced. This is a size effect by its origin that directly depends on the extent of binding of nanoparticles with DNA molecules. Therefore, the conformation homogeneity of DNA molecules can be distorted under the effect of Au nanoparticles of a specific size only, which can be followed by changes in the abnormal optical activity of CLCD.

The considered facts allow one to add a few comments regarding Figure 4.22. First of all, the amplitude of the abnormal band in the CD spectrum depends mainly on helical twisting of quasinematic layers in cholesteric structures of CLCDs [60, 68].

Second, the theory developed by Belyakov et al. [39] predicts that the change in the efficiency of interaction between molecules in quasinematic layers must be accompanied by changes in the parameters of the spatial helical structure of CLCD particles and in the value of amplitude of an abnormal band in the CD spectra. This means that the decrease in the amplitudes of abnormal bands in the CD spectra (curves 2–4 and 2^I–5^I) is connected with the change in the extent of helical twisting of quasinematic layers in the structure of CLCDs formed by any type of nucleic molecules, i.e., B-family for DNA and A-family for poly(I)×poly(C). This effect reflects only one process, i.e., an incorporation of Au nanoparticles into content of CLCD particles formed by these molecules. Third, any possible aggregation of independent Au nanoparticles outside of CLCD particles cannot induce a change in the real value of the abnormal band in the CD spectra in the region of absorption of biopolymeric molecules.

The efficiency of the reduction in the amplitudes of abnormal bands for the CLCDs formed by DNA or poly(I)×poly(C) molecules (Figure 4.22) depends on the size of the Au nanoparticles and the time of treatment with Au nanoparticles. For instance, if the diameter of the Au nanoparticles is 2 nm, the amplitude of the abnormal band decreases by 75% after 4 hours of treatment, but in the case of 15-nm Au nanoparticles, it decreases by only 20% [69]. This fact allows one to suggest that the incorporation of Au nanoparticles in the content of CLCD is determined by the distance between nucleic acid molecules packed in the quasinematic layers [60]. Au nanoparticles with a size of about 2 nm are capable of diffusing between DNA molecules, because this size is close enough to the distance between DNA molecules obtained under the given experimental conditions (concentration of PEG, etc.) [69]. However, Au nanoparticles with sizes of 5 nm and 15 nm are too big in comparison to the distance between nucleic acid molecules in quasinematic layers. Hence these particles do not diffuse effectively into the content of the formed CLCD particles.

Considering the results shown in Figure 4.22, one can take into account the following. The first experiments [70, 71] demonstrated that Au nanoparticles can be assembled near DNA molecules forming supramolecular structures. Later, it was shown that an assembly of the Au nanoparticles is accompanied by formation of a planar superstructure consisting of repetitive neighboring linear DNA molecules and Au nanoparticles. These results unequivocally show that linear, rigid DNA molecules interact with Au nanoparticles to form planar superstructures of type (...-Au-DNA-Au-DNA-Au-DNA-Au-...) despite the anisotropic properties of the initial DNA molecules [64, 65, 72–74].

Hence, on the one hand, Au nanoparticles enforce DNA molecules to be organized into planar superstructures, in which neighboring molecules are not only closely packed, but collaterally located. On the other hand, it is well known that neighboring DNA molecules packed in quasinematic layers of CLCD particles are helically twisted due to the anisotropic properties of these molecules [60].

Comparison of these two circumstances means that immobilization (incorporation) of Au nanoparticles in quasinematic layers will, indeed, result in untwisting of neighboring layers. This explains the changes in the CD spectra shown in Figure 4.22. Indeed, the steric limitations resulting from both dense packing of nucleic acid molecules in quasinematic layers and fixed PEG concentration in solution prevent the spatial separation (denaturation) of the two strands of the neighboring molecules [60]. In this case, both the incorporation of Au nanoparticles into quasinematic layers and their ordering near surfaces of DNA molecules must be transformed into the changes in the energy of interaction between neighboring molecules in quasinematic layers.

Hence, Au nanoparticles can induce a transition from a cholesteric structure of CLCD particles that has high abnormal optical activity to a certain structure having very low (if any!) abnormal optical activity. For this process, the differences in the secondary structure of various nucleic acid molecules are not important. This means that the changes in the CD spectra clearly demonstrate that Au nanoparticles are incorporated within the structure of CLCD particles. These changes in the CD spectra depend only on the presence of Au nanoparticles in the content of quasinematic layers. The concrete peculiarities of the mechanism of interaction of Au nanoparticles with nucleic acid molecules are not important for this effect.

Additional study of visible absorption spectra of Au nanoparticles provides a way to evaluate the size of the assemblies formed by these particles under various conditions [75–82]. Figure 4.23(A) shows the time-dependent changes in the visible optical spectrum that were observed during the treatment of the DNA CLCDs with Au nanoparticles. (In Figure 4.23(A), curve 1 is for 5 min; curve 2 is for 12 min; curve 3 is for 43 min; curve 4 is for 109 min; curve 5 is for 230 min; curve 6 is for 1200 min after the addition of Au nanoparticles. Conditions: $C_{DNA} = 9$ μg/mL; $C_{PEG} = 150$ mg/mL; 0.27 mol/L NaCl + 1.78×10^{-3} mol/L phosphate buffer; $C_{nano-Au} = 0.82 \times 10^{14}$ particle/mL. In Figure 4.23(B), curve 1 (open squares) depicts experimental data from Figure 4.23(A); curve 2 (filled squares) depicts data from Link and El-Sayed [77]; curve 3 (filled circles) depicts data from Khlebtsov et al. [81]; curve 4 (triangles) depicts data from Dykman et al. [82].)

First of all, the treatment is accompanied by an appearance of absorption at $\lambda \approx 505$ nm (surface plasmon resonance [SPR] absorption band) [75, 77]. Similar results for double-stranded poly(I)×poly(C) are not shown.) The plasmon resonance is responsible for the pale-violet color of solution containing DNA CLCD treated with Au nanoparticles. This resonance is definitely absent in the initial DNA solution and only slightly defined at $\lambda \approx 505$ nm in colloid solution containing Au nanoparticles with a size of about 2 nm. This means that the immobilization of Au nanoparticles in the content of DNA CLCD particles (Figure 4.23) brings Au nanoparticles close together. This results in modifying their local environment and changing the position of the SPR band, because the SPR absorbance is sensitive to the local environment [80]. Secondly, the intensity of the SPR band is gradually increased, and the maximum of this band is shifted from $\lambda \approx 505$ nm to $\lambda \approx 550$ nm over time.

Comparison of data (Figure 4.23(B), curve 1) to available numerical data from the literature [75–79, 81, 82] concerning the correlation between the position of the maximum of the SPR band and the size of Au nanoparticles shows that the change in the SPR band reflects the increase in the size of Au assemblies from 2 to about 50 nm. Of course, this is a very simple estimation that does not take into account such factors as the real shape of Au assemblies, the local dielectric constant, or the local refractive index [83].

The comparison of the SPR band typical for PEG containing water–salt solutions added with 5-nm Au nanoparticles to an analogous solution containing CLCDs

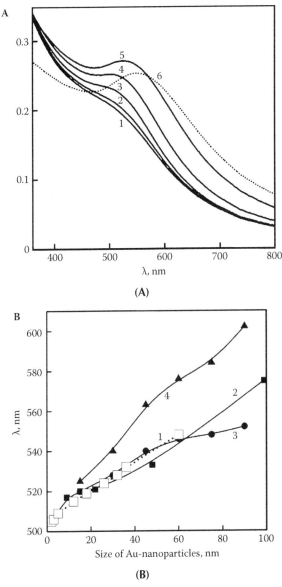

FIGURE 4.23 Change in the absorption spectrum of Au nanoparticles (2 nm) at (A) their interaction with double-stranded DNA CLCD and (B) the correlation between the position of SPR band and the size of Au nanoparticles.

formed by DNA molecules showed no changes in the SPR band amplitude. The absence of changes in the SPR band was observed in the case of CLCD treated with 15-nm Au nanoparticles as well (data are not shown). These results confirm once again that big Au nanoparticles do not incorporate into the structure of quasinematic layers of CLCD particles.

Here one can add that due to the spherical symmetry of the individual Au nanoparticles used by us, the formation of assemblies from these particles is a significant problem. But Figure 4.23 shows that the DNA molecules, ordered in the quasinematic layers, provide, as templates, the necessary symmetry-breaking mechanism to form extended linear structures from Au nanoparticles fixed between neighboring double-stranded DNA molecules. Finally, after twenty hours, one can detect the formation of a dark-violet precipitate in the tested solution. Intensive stirring of the solution containing the pellet restores its optical properties, which makes it possible to measure its optical spectrum. Curve 6 in Figure 4.23(A) demonstrates the presence of a plasmon resonance band located at $\lambda \approx 545$–550 nm. Close similarities between the shapes and positions of the SPR bands (curves 6 and 5) show that the dark-violet precipitate consists of isolated DNA CLCD particles, which, in turn, contain extended structures (about 50 nm) formed by Au nanoparticles. This result shows that the spectral changes in the visible optical spectrum are induced by the processes that take place, mainly, within isolated CLCD particles treated with Au nanoparticles, but not between neighboring particles.

The diminishing amplitudes of abnormal bands in the CD spectra and the increase in the SPR bands for DNA and poly(I)×poly(C) CLCDs treated with Au nanoparticles showed that there are two different processes:

1. A quick drop in the amplitudes of the CD bands due to incorporation of Au nanoparticles into quasinematic layers of CLCD particles
2. A relatively slow shift of the SPR band due to the assembly of neighboring Au nanoparticles (formation of Au clusters) in the content of CLCD particles

These two independent processes start simultaneously but differ in their velocities. Hence, the comparison of Figure 4.22 to Figure 4.23 speaks in favor not only of incorporation of Au nanoparticles, but formation of Au clusters, growing with time [75–79], in the content of CLCD particles.

At fixed osmotic pressure and ionic strength of a PEG-containing solution, the formation of an insoluble pellet (see previous discussion) shows that Au clusters, located in quasinematic layers and interacting (independently on mechanism) by this or that way with nucleic acid molecules results in a decrease in the solubility of these molecules. In this case, the transition of whole CLCD particle into an insoluble (rigid) state takes place. Moreover, the formation of a multilayer sandwich-like structure consisting of alternate Au clusters located both between DNA molecules and in neighboring quasinematic layers would lead to an additional contribution to the energy of stabilization of these rigid CLCD particles [73, 84].

One can stress that the transition of CLCD into a rigid (insoluble) state takes place at a definite concentration of Au clusters. Hence, the interaction of Au nanoparticles (2 nm) with neighboring DNA molecules leads to the formation of a spatially fixed structure of the CLCD. The stability of this structure is determined by the number and properties of Au clusters rather than the properties of the initial PEG-containing solution. This means that a new structure would persist even in the absence of the high osmotic pressure of the solution. In this case, it appears that there is a possibility

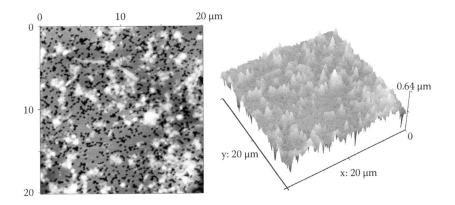

FIGURE 4.24 AFM images of the DNA CLCD particles treated with Au nanoparticles and immobilized on the surface of a nuclear membrane filter. (The dark spots are holes in the nuclear membrane filter.)

to investigate the properties of a newly formed structure. For instance, the insoluble CLCD particles formed by [DNA-Au] complexes can be immobilized on the surface of a nuclear membrane filter and the AFM images of these particles can be registered.

As an example, Figure 4.24 demonstrates the images (2-D and 3-D) of DNA CLCD particles treated with 0.82×10^{14} Au nanoparticles (2 nm) per mL. (In Figure 4.24, the left panel is a 2-D image and the right panel is a 3-D image. Conditions: $C_{\text{Nano-Au}} = 0.82 \times 10^{14}$ particle/mL.) The shape of the particles is close to elongated spheroids. One can stress that this shape is very similar to the shapes of particles formed as a result of formation of nanobridges between neighboring DNA molecules, or as a result of the decrease in solubility of DNA treated with the salt of REE [85]. In the case here, the sizes of particles are varied from 0.1–0.2 μm to 0.7 μm, with 0.4–0.5 μm as the average. This means that the initial size of the CLCD particles was not practically changed at interaction of Au particles with DNA.

Figure 4.24 confirms as well that DNA CLCD particles treated with Au particles exist as independent, insoluble objects. The presence of single particles (Figure 4.24) demonstrates that upon treatment of DNA CLCD by Au nanoparticles, the liquid character of the DNA packing in the particles disappears, and the formed particles have a rigid spatial structure. Moreover, the large sizes of rigid CLCD particles formed by [DNA-Au] complexes speak against coalescence of Au nanoparticles outside CLCD particles.

Results obtained using a confocal microscope indicate that the CLCDs of [DNA-Au] complexes exist as independent objects. The CLCD [DNA-Au] complexes formed in PEG-containing solution (170 mg/mL of PEG) were additionally treated with SYBR Green dye. The fluorescence images (Figure 4.25) obtained by the confocal microscope demonstrate that particles of CLCD of [DNA-Au] complexes exist as isolated objects. (Conditions for Figure 4.25: $C_{\text{DNA}} = 9$ μg/mL; $C_{\text{PEG}} = 150$ mg/mL; 0.27 mol/L NaCl + 1.78×10^{-3} mol/L phosphate buffer; $C_{\text{Nano-Au}} = 0.82 \times 10^{14}$ particle/mL; $C_{\text{SYBR Green}} = 4.11 \times 10^{-6}$ mol/L.)

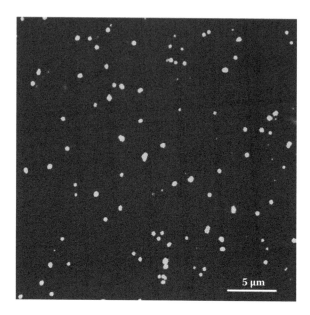

FIGURE 4.25 **(See color insert)** Fluorescence image of CLCD particles formed by DNA in PEG-containing solution and treated with Au nanoparticles (2 nm) and then with SYBR Green.

The mean fluorescence "diameter" of these particles is similar to the mean diameter detected for initial DNA CLCD particles treated as well with SYBR Green. This indicates no strong deformation of the DNA secondary structure due to interaction with the Au nanoparticles. The data shown in Figure 4.25 are important. They confirm that an appearance of the SPR band is connected with the assembly of Au nanoparticles within individual particles of CLCD, in other words, with formation of Au clusters between neighboring DNA molecules in quasinematic layers of CLCD.

The application of small-angle X-ray scattering (SAXS) to the investigation of phases formed by DNA CLCD particles treated with Au nanoparticles was used to obtain new results related to a possible model of these particles. First of all, it was shown that the distance *between* neighboring quasinematic layers is not changed as a result of incorporation of Au nanoparticles into the structure of DNA CLCD particles. Second, the distortion of ordering of DNA molecules in neighboring quasinematic layers in CLCD particles is increased. This result correlates with the CD spectra shown in Figure 4.22, which gives evidence for a decrease in spatial ordering of DNA quasinematic layers when concentration of Au nanoparticles is increased. Hence, one can repeat again that the incorporation of Au nanoparticles into the structure of quasinematic layers of DNA CLCD particles enforces the collinear location of DNA molecules, i.e., this process is accompanied by untwisting of neighboring layers. Finally, samples containing Au nanoparticles are strongly scattering at very small angles, i.e., in the region of the central scattering, which points to the presence of sufficiently large polydisperse scattering objects (clusters of Au nanoparticles).

The viewpoint about the possible linear arrangement of Au nanoparticles and the results of size calculation of Au clusters made by the program GNOM [86] allows one to suggest a hypothetical model reflecting the incorporation of Au nanoparticles between DNA molecules in quasinematic layers (Figure 4.26).

According to this model, initial Au nanoparticles have an average diameter (size) of about 2 nm. These nanoparticles form linear clusters with an average length of 13–15 nm and clusters with a maximal length of ≈40 nm. Formation of Au clusters between DNA molecules induces alteration of mutual orientation of neighboring quasinematic layers. Under these conditions, the helical twisting of these layers in the structure of CLCD particles is decreased. Thus, the results of SAXS support the previous statement, according to which the decrease in amplitude of an abnormal band in the CD spectrum of DNA CLCD treated with Au nanoparticles reflects the distortion in the arrangement of neighboring quasinematic layers induced by formation of the Au clusters. The particles lose their helical twisting, and a transition of type "helically twisted structure → structure without twisting" is induced. Moreover, the formation of Au cross-links connecting neighboring DNA molecules is possible. Due to the real position of these molecules in quasinematic layers, one can expect that, in order to connect these molecules, the ends of an Au cross-link should wrap around the double-stranded DNA molecules [65]. It is not excluded that Au cross-links can induce an additional stabilization of the rigid structure of DNA CLCD particles.

Hence, the obtained results demonstrate that interaction of Au nanoparticles with DNA CLCD results in a set of effects:

1. It facilitates reorganization of the initial cholesteric structure.
2. It induces the formation of Au clusters within particles of DNA CLCD.
3. It induces the transition from a liquid to a rigid structure of DNA CLCD particles, i.e., the formation of a new type of rigid DNA nanoconstruction.

These results focus attention on the biological "fate" of Au nanoparticles in living cells. Indeed, the incorporation of Au nanoparticles in cells has been investigated in

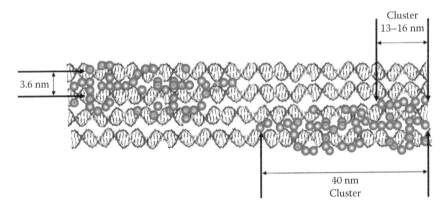

FIGURE 4.26 (See color insert) Hypothetical model of location of Au clusters between double-stranded DNA molecules forming the quasinematic layer.

many laboratories [87–90]. However, little is known about the effect of Au nanoparticles at the level of a cell nucleus. (The cell nucleus realizes all processes that occur within the cell, and any disruption within the nucleus would subsequently affect the cell's DNA, thereby disturbing the highly regulated cell cycle.) Here we can recall that DNA CLCD particles reflect some properties of protozoan chromosomes. Induced by small Au nanoparticles, changes in the spatial structure of DNA CLCDs allow one to suggest that any of the considered effects due to possible absorption of Au nanoparticles between DNA fragments of the cell chromosomes can be the reason for the genotoxicity of these particles [89, 91].

4.9 AN ADDITIONAL CASE: FORMATION OF SEMI-RIGID LIQUID-CRYSTALLINE PARTICLES

It was shown by Yevdokimov et al. [60] that there is a DNA phase exclusion from water–salt solutions as a result of the "correlation attraction" [92, 93] between the segments on nucleic acids, whose phosphate group negative charges are neutralized with positively charged polycations.

Since the moving force of this process is the change in the enthalpy of the system, this method of phase exclusion is called "enthalpy condensation." Polycations that neutralize a large number (80%–90%) of negative charges of DNA phosphate groups are used to realize this method. Then the attraction between the DNA molecules becomes strong enough to induce a spontaneous condensation. The attraction forces at the condensation are mostly electrostatic (electrodynamic) by nature, including the London dispersion forces and dipole-induced dipole interaction. These forces, which are relatively small at long distances between DNA molecules, increase rapidly as the DNA molecules approach one another, while the attraction energy increases as $1/r^5$. The value of the dispersion force is proportional to the Hamaker empiric constant, A, which is $\approx 4 \times 10^{-14}$ erg for organic molecules interacting in water, that is, about 1 kT at room temperature. For the attraction at a distance of about 3.0 nm, the value of A varies from 2 to 5 kT.

It is obvious that, when the surface charge density of DNA molecules decreases to a certain extent (in particular, because of the neutralization of their phosphate group negative charges by positive charges of polycations added to the solution), the dispersion forces balance and then exceed the electrostatic repulsion of neighboring molecules of [DNA-polycation] complexes. In this case, the correlation attraction [93] between almost-neutral molecules causes the phase exclusion of the formed [DNA-polycation] complexes from the solution, i.e., the formation of dispersion, whose particles can be characterized by an ordered (liquid-crystalline) packing of molecules of [DNA-polycation] complexes. Polyamines, polyamino acids, proteins (histones), and so forth, were used as polycations, causing the formation of LCD [94–97].

The properties of [DNA-polycation] dispersion particles usually differ from the properties of DNA LCD formed by entropy condensation. These differences consist of the following:

First, polycation molecules fixed on the surface of DNA molecules are always entered into the content of [DNA-polycation] LCD particles formed in water–salt solutions. They can play the role of a dielectric medium, changing the mode of interaction between the DNA molecules.

Second, the phase exclusion of molecules of [DNA-polycation] complexes takes place at the critical concentration of the polycation in the solution, determined by the value of the binding constant of polycation to the DNA. The higher the binding constant, the lower the critical concentration of polycation, which causes the phase exclusion; at the same time, the value of the binding constant of the polycation is the function of its molecular mass.

Third, considering the fact that the polycation molecules have a fixed spatial conformation, the energy of the interaction between the [DNA-polycation] molecules provides a constant (fixed) distance between neighboring molecules of complexes.

Fourth, the combination of the first and the third points shows that the polycations play a dual role: On the one hand, when interacting with the DNA, they change the homogeneity of the charge distribution on the surface of these molecules; on the other hand, they act as a medium that modifies the efficiency of interaction between the neighboring DNA molecules. Under these conditions, the type of "recognition" of neighboring [DNA-polycation] molecules is different from the recognition of the initial anisotropic DNA molecules. The molecules of [DNA-polycation] complexes are ordered so that, as a rule, a hexagonal (or very close to hexagonal), not cholesteric, packing is realized. Such molecule packing corresponds to the Bragg distances within the limit of 2.6–2.9 nm, which are typical of the hexagonal packing of double-stranded DNA molecules.

Fifth, depending on the empirical combination of several factors, notably, the ionic strength of the solution, the spatial structure of the polycation molecules, the content of positively charged groups, etc., only in some cases is it possible to form an LCD characterized by cholesteric [DNA-polycation] molecules packing. In these cases, the abnormal optical activity typical of the [DNA-polycation] complexes of CLCD particles makes it possible to observe the changes in the mode of packing of the molecules. In particular, in the case of interaction of calf thymus DNA with poly-L-lysine and its analogues [98], the formation of dispersions with different signs of low-intense bands in the CD spectra, which is evidence of the formation of CLCD particles with a different sense of helical twist of the structure of the dispersion particles, was observed.

The study of the enthalpy condensation of DNA molecules induced by their interaction with linear, nontoxic, biocompatible, biodegradable poly(aminosaccharide)-chitosan (copolymer consisting of residues of β-(1→4)-2-amino-2-deoxy-D-glucopyranose and β-(1→4)-2-acetamido-2-deoxy-D-glucopyranose) attracts a great deal of interest from physicochemical and biological points of view (Figure 4.27).

The chitosan (Chi) molecule has positively charged amino groups and can form complexes with DNA molecules because its phosphate groups carry negative

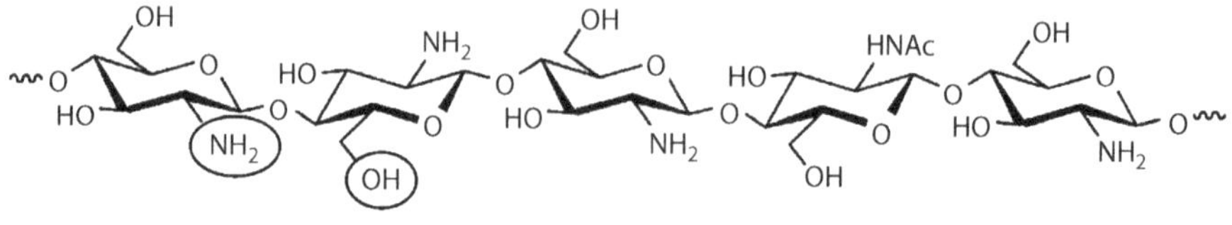

FIGURE 4.27 Structural formula of a fragment of a chitosan molecule. (The chemical groups capable of forming chelate complexes with metal ions are marked.)

charges. All details of this process were analyzed by Yevdokimov et al. [60]. Thus we only outline the most important peculiarities of the interaction between Chi and DNA molecules.

It should be noted that the formation of complexes between macromolecules is significantly different by its parameters from the formation of complexes between low-molecular-mass compounds. This difference is caused by *polymeric effects*, which induce the formation of relatively stable complexes even when the energy of interaction between macromolecules is low. The efficiency of the interaction between macromolecules is a function of the extent of polymerization, and the apparent equilibrium constant increases sharply at a certain critical length of the polymeric chain. The fact that D-glycosamine (the main component of Chi) does not form an insoluble complex with DNA and, in addition, the efficiency of the formation of DNA LCD significantly depends on the molecular mass of Chi [99] speaks in favor of what was stated previously. Beginning with a length of Chi molecule that is equivalent to twenty Chi monomer units, i.e., starting with the length of about 10 nm [100, 101], Chi cooperatively interacts with DNA molecules. This evaluation does not consider the fact that at selected conditions, only 50% of Chi amides can interact with DNA and, moreover, due to the peculiarities of the spatial structure of Chi molecules [100, 101] only the every-other-one of Chi amino group can interact with the DNA negatively charged phosphate groups. The result means, in its turn, that a minimal fragment of a double-stranded DNA molecule that can form an insoluble complex with Chi must contain at least thirty base pairs (10 nm/0.34 nm), which corresponds well to the results of the evaluations (fifteen–twenty pairs) by other authors [102].

Chi molecules interact with DNA so that the amino groups of sugar residues both neutralize the negative charges of DNA phosphate groups and create a specific distribution of positively charged molecules of the DNA surface. In this case, Chi molecules bonded to the DNA molecules play the role of a dielectric medium that affects the mode and the efficiency of interaction between neighboring DNA molecules at the "moment of their approaching" necessary for the formation of particles of dispersion. It means that a significant contribution to the efficiency of the interaction between Chi and DNA must be made by the conformation of the carbohydrate chain of a Chi molecule.

The efficiency of interaction of Chi with DNA molecules [60] depends upon standard factors such as ionic strength, pH value of solution, etc. However, the peculiarities of the spatial structure of Chi molecules play here a very noticeable role. The Chi amino groups can interact with DNA phosphate groups only by the next-to-nearest type [100, 101]. This leads to the fact that, when interacting with the DNA, Chi molecules

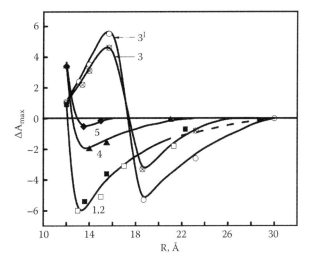

FIGURE 4.28 Dependence of the maximum amplitude of the band ($\lambda = 270$ nm) in the CD spectra of CLCDs formed by [DNA-Chi] complexes upon the average distance (R) between amino groups in Chi molecules.

are arranged along the long axis of a DNA molecule and form a helical structure from the Chi dipoles. At a certain distance between the charged amino groups of Chi groups located along the DNA, the corresponding contribution into the dispersion interaction between neighboring molecules of the initial DNA due to dipole-dipole interaction may appear. In this case, the sense of the spatial twist of cholesteric helix existing in the [DNA-Chi] complex CLCD particles may change, which, in its turn, will lead to the change of the sign of the abnormal optical activity of the formed dispersion.

Figure 4.28 shows the dependence of the maximum amplitude, ΔA_{max}, of the band in the CD spectra of LCDs formed by [DNA-Chi] complexes on the average distance, R, between amino groups in the Chi molecules. (In Figure 4.28, the numbers above the curves are the NaCl concentrations in the solutions where the [DNA-Chi] complexes were formed; curves 1, 4, and 5 depict the Chi molecular mass of 14,600 g/mol at 0.05, 0.15, and 0.5 mol/L NaCl, respectively; curve 2 is for Chi molecular mass 8,400 g/mol at 0.05 mol/L NaCl; curve 3 is for Chi molecular mass 5,000 g/mol at 0.05 mol/L NaCl; curve 3 turns into curve 3′ when considering the Chi concentration that is not bound in complexes with DNA. Conditions: $C_{DNA} = 15.5$ µg/mL; 0.001 mol/L phosphate buffer; pH 6.86; $\Delta A = (A_L - A_R) \times 10^{-3}$ optical units.)

The curves in Figure 4.28 show that the contribution of two parameters—the ionic strength of the solution and the percentage of amino groups—determines the possibility of the formation of two types of CLCDs by [DNA-Chi] complexes, characterized by either negative or positive abnormal bands in the CD spectra. One can see that the sign of the abnormal band in the CD spectrum of [DNA-Chi] complex CLCD changes with a decrease of the distance between charged amino groups in a Chi molecule. This means that the introduction of an extra number of charged groups into the structure of the complex (for instance, by intercalation of some substances between DNA base pairs) must affect the value and even the sign of the abnormal band in the CD spectrum of CLCD of [DNA-Chi] complexes. Moreover, under certain conditions, [DNA-Chi] complexes could form LCDs having no abnormal band in the CD spectrum, and ΔA_{max} becomes zero at both long and short distances between amino groups in the Chi molecules. This means that the properties of Chi molecules (molecular mass, charge distribution, etc.) can control the abnormal optical activity of [DNA-Chi] LCDs.

Therefore, the Chi example shows that the spatial structure of [DNA-polycation] LCD particles is regulated by a complicated combination of such factors as the conformation of the polycation molecule, the distance between the positive charges in the polymer chain, the ionic strength, the pH of solutions, etc. This means that the enthalpy condensation of double-stranded DNA molecules with low-molecular-mass depends on a combination of multiple factors, and only at a certain combination of polycation properties and properties of the solvent, can CLCD be formed from [DNA-polycation] complexes.

One can stress that the Chi molecules in [DNA-Chi] complexes bring a new reactivity and open the possibility of modifying a whole system. Since neighboring free amino and hydroxyl groups in Chi molecules effectively form chelate complexes with copper ions [103–108], the attempt to induce cross-linking of neighboring Chi molecules fixed in the structure of CLCD particles of [DNA-Chi] complexes, i.e., an attempt to create a new type of nanoconstruction, has been performed [60]. The formation of nanobridges from the alternation of DAU molecules and copper ions between neighboring Chi molecules fixed near DNA molecules results in a very interesting conclusion. A certain conformation of Chi molecules is needed for nanodesign. Moreover, there is a minimal distance between amino groups (about 1.3 nm, which approximately corresponds to the every-other-one distance) at which neighboring [DAU-Cu^{2+}] nanobridges can be fixed. For this case, the low abnormal optical activity is typical of CLCD particles of [DNA-Chi] complexes treated by DAU and copper solutions. Only with an increase in the distance between amino groups in Chi molecules can the steric structure of the Chi ensure the orientation of amino and hydroxyl groups in sugar residues that is appropriate for formation of optimal nanobridges in the nanoconstruction formed. This nanoconstruction contains molecules, not only DNA, but guests such as Chi and DAU molecules as well as copper ions. Investigation of the properties of this new type of [DNA-Chi] nanoconstruction is now in progress.

The effect of another specific factor, namely, the diameter of the polycation molecule, on the ability of [DNA-polycation] complexes to form CLCD particles has been estimated as well. From this point of view, the properties of dispersions that formed at the interaction between DNA molecules and dendrimers—branched polycations with definite size, high density of the surface charge, and sufficient commercial availability—attract special interest. Several research teams [109–111] have attempted to study the interactions of dendrimers with DNA and demonstrated its electrostatic nature, with the efficiency depending on solution ionic strength. Taking into account that high-generation dendrimer molecules have a voluminous spatial structure, the role of the size of the dendrimer molecule as an additional factor affecting the packing mode of DNA molecules was analyzed.

The interaction between poly(propyleneimine) dendrimers (DAB-dendr-$(NH_2)_x$) of five generations (G1–G5) and molecules of double-stranded DNA in water–salt solutions with different ionic strength (0.01–0.7) was described in detail by Skuridin et al. [112].

DAB-dendr-$(NH_2)_x$ are water-soluble tree-like polymeric molecules with a branched spatial structure (Figure 4.29). At physiological pH values, DAB-dendr-$(NH_2)_x$ molecules carry a positive charge, which determines their affinity for polyanionic compounds, in particular for DNA. The interaction between DAB-dendr-$(NH_2)_x$ molecules and DNA gives poorly soluble complexes [113].

The most interesting results have been highlighted by Skuridin et al. [112]. First, it has been proven that regardless of the ionic strength of the solutions and the structure of the dendrimer, the bonding of DAB-dendr-$(NH_2)_x$ molecules to DNA leads to the formation of a dispersion. By means of circular dichroism, the border conditions (ionic strength of the solution and the molecular mass of DAB-dendr-$(NH_2)_x$ necessary for the formation of optically active and optically inactive dispersions of [DNA-DAB-dendr-$(NH_2)_x$] complexes were evaluated. It was found that in the case of DAB-dendr-$(NH_2)_x$ molecules of the first, the second, the third, and the fifth generations, the dispersions are formed, but they do not have an abnormal optical activity. The situation in the case of the fourth-generation dendrimers is different. The interaction of the DNA molecules with DAB-dendr-$(NH_2)_x$ of the fourth generation results in formation of two types of LCD.

Figure 4.30 shows the CD spectra of DNA water–salt solutions registered at different concentrations of G4 in the case of the high ionic strength of solutions ($\mu > 0.4$). (In Figure 4.30, curve 1 is for no G4; curves 2–6 are for 4.82, 9.61, 11.99, 12.95, and 13.90×10^{-5} mol/L of G4, respectively. Conditions: $C_{DNA} = 32.31$ µg/ml; 0.6 mol/L NaCl + 10^{-2} mol/L phosphate buffer, pH ≈ 7.0; G4 molecular mass = 3,060 Da; $\Delta A = (A_L - A_R) \times 10^{-3}$ optical units.)

It can be seen that the CD spectrum of the initial linear B-form DNA solution (curve 1) changes when G4 concentrations exceed a given C^{cr} value and a negative band located in the absorption region of nitrogen bases appears, and its amplitude reaches its maximum ($\Delta A_{max} \approx -6,200 \times 10^{-6}$ optical units) at G4 = 13.90×10^{-5} M (curve 6). As the amplitude of the negative CD band of the [DNA-G4] dispersion considerably exceeds the ΔA value typical of the molecular optical activity of nitrogen bases in isolated DNA molecules ($\Delta A = 230 \times 10^{-6}$ optical units; compare curves 1 and 6 in Figure 4.30), it can be stated that the [DNA-G4] dispersion displays an abnormal optical activity. However, in solutions with low and moderate ionic force ($\mu < 0.4$), dispersions with no intense negative band in their CD spectra are formed.

Theoretically calculated CD spectra of DNA LCDs [60] demonstrate that the intense (abnormal) band in the spectra in the absorption region of nitrogen bases of this macromolecule is associated with two parameters: helical packing of DNA molecule and a high local density of chromophores (nitrogen bases). These conditions are fulfilled in the case of a liquid-crystalline cholesteric packing of DNA molecules in dispersion particles. Hence, the data in Figure 4.30 allow one to state that the spatial structure is typical of particles of [DNA-G4] complexes. The fact that the [DNA-G4] dispersion displays an intense negative CD band demonstrates that, in this case (as well as in the

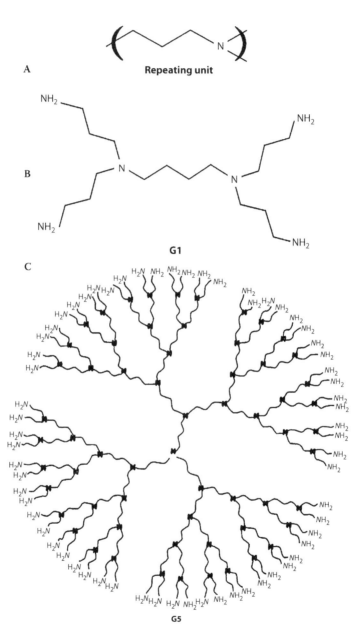

FIGURE 4.29   Schematic structures of diaminobutane (A), repeated structural unit in poly-propylenimine dendrimers, and the molecules of these compounds of first generation (G1; B) and fifth generation (G5; C).

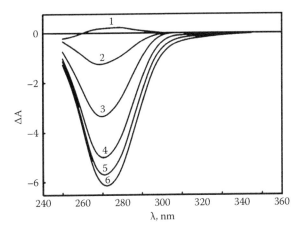

FIGURE 4.30 Changes in the CD spectra of DNA water–salt solutions induced by different G4 concentrations and formation of dispersion.

case of classical DNA cholesteric structures), the molecules of [DNA-G4] complexes have a left-handed cholesteric helical twist in dispersion particles.

Two facts should be taken into account when considering the reasons that determine the stability of the particles formed by complexes and their optical properties [114]. First, the common feature for colloids and emulsions is that their structural stability is of a kinetic rather than of a thermodynamic nature. Note that some emulsions have a short lifetime before a complete phase separation, whereas others can remain stable for a long time. To stabilize an emulsion, it is supplemented with a surfactant, which preserves its drops from immediate coalescence. In addition, particles can be stabilized by increasing their surface charge. It cannot be excluded that in the case of G4, when only one side of these molecules (due to steric reasons) can interact with a DNA molecule, the formed particles of complexes carry an additional surface charge. Indeed, the positive charges on the surface of G4 molecules are concentrated in a rather small area in comparison to the negative charges on the surface of DNA molecules. Thus, the steric limitations provide for an additional positive charge of the molecules of [DNA-G4] complexes. Under these conditions, the particles of complexes themselves will display an increased stability. Second, as only one side of a rigid linear G4 molecule is occupied by a DNA molecule, the other side, which also carries a large number of positive charges, can also serve as a binding site for another DNA molecule. This process can spread over large distances.

Considering the role of the size (diameter) of DAB-dendr-$(NH_2)_x$ molecules as a factor influencing the interaction efficiency of these compounds with double-stranded DNA molecules, the following should be noted:

1. The correlation between the diameter of poly(amidoamine) dendrimer molecules and their generation number has a universal character [112]. This statement also applies to poly(amidoamine)-based dendrimers, namely, for G4 and G5 molecules with a diameter of approximately 3.12 nm and 3.96 nm, respectively [109], and these data correspond to the universal curve.

2. Despite an electrostatic interaction (repulsion) of DNA molecules carrying a negative charge, these molecules can precipitate from solutions in the presence of multivalent ions. According to theoretical estimates, such precipitation is determined by an efficient short-range interaction (attraction) between the fragments of neighboring DNA molecules.

The precipitation of polyelectrolyte molecules is theoretically explained taking into account the specific features of the model [112], postulating that polyelectrolyte contains the fragments of two types, which appear as a result of ion condensation. The first-type fragments carry condensed ions, whereas the second-type fragments retain their initial negative charge. Owing to a strong interaction with the surface of polyelectrolyte molecules and one another, condensed ions are not randomly distributed in a 3-D space, but rather form a strongly correlated system on the surface of polyelectrolyte molecules. In addition, a charge inversion of polyelectrolyte molecules becomes possible. The total short-range electrostatic attraction between different types of fragments in neighboring polyelectrolyte molecules determines their precipitation. In this process, the higher the valence of condensed ions, the stronger is the attraction between the fragments of different types. Therefore, it can be expected that when Na^+ ions are replaced with multiply charged cations, DNA molecules will precipitate.

A fundamental component of all models explaining the attraction between polyelectrolyte molecules is the role of condensed counterions in this process. The attraction between DNA molecules is a short-range interaction. This means that the electrostatic repulsion determined by the presence of uncompensated charges is predominant at larger distances, i.e., that two DNA molecules will repulse from one another. Only in the case when two DNA molecules are close enough to one another, the attraction caused by the correlation will determine a parallel arrangement of these molecules [93, 115]. The correlation between condensed ions on the neighboring DNA molecules underlies the mutual attraction of these macromolecules [116].

3. The fact that DNA molecules display geometric and optical anisotropies means that the attraction between these macromolecules will place them at a certain angle to each other, leading to a cholesteric spatial structure in the particles of the formed dispersion.

Being located on the surface of DNA molecules, spherical DAB-dendr-$(NH_2)_x$ molecules will perform two functions in the attraction of the latter. They will play the role, first, of the medium determining the attraction efficiency between neighboring DNA molecules and, second, of the joints allowing the neighboring DNA molecules to rotate. It is evident that the diameter of DAB-dendr-$(NH_2)_x$ molecules will play an important role in both cases.

Precipitation (phase separation) of DNA molecules takes place at a certain extent of negative charge neutralization in these molecules by the positive charges of DAB-dendr-$(NH_2)_x$ molecules. These conditions are provided at a critical DAB-dendr-$(NH_2)_x$ concentration in solution: The DAB-dendr-$(NH_2)_x$ complex becomes electrically neutral [109], and in this state the particles of a [DNA-DAB-dendr-$(NH_2)_x$] dispersion precipitate. Indeed, at C_{G4} below $\approx 6.0 \times 10^{-5}$ mol/L, the [DNA-G4] complex is in a soluble form, and at $C_{G4} > \approx 8.0 \times 10^{-5}$ mol/L the sedimentation takes place, which is accompanied by changes in its absorption and CD spectra.

Taking these facts into account, the data shown in Figure 4.31 attract interest. (In Figure 4.31, $C_{DNA} \approx 30$ µg/mL; 10^{-2} mol/L phosphate buffer; pH ≈ 7.0; and $\Delta A = (A_L - A_R) \times 10^{-3}$ optical units.) It is evident that the largest negative band amplitude in the case of dispersions of [DNA-DAB-dendr-$(NH_2)_x$] complexes corresponds to G4, i.e., when the distance between neighboring DNA molecules determined by the diameter of G4 molecules is ≈ 3.12 nm. With the further increase in diameter of the DAB-dendr-$(NH_2)_x$ molecule, the CD band amplitude of [DNA-DAB-dendr-$(NH_2)_x$] particles decreases. This decrease is not connected with the dissolution of the [DNA-DAB-dendr-$(NH_2)_x$] complex but, rather, reflects a decrease in the interaction efficiency between DNA molecules, which is necessary for their helical twisting. The decrease in the CD band amplitude of [DNA-DAB-dendr-$(NH_2)_x$] complexes can be caused by two reasons. First, this can be a consequence of a complicated dependence between the band amplitude and the distance between the amino groups in the DAB-dendr-$(NH_2)_x$ molecule [99], which shields the negative charges of DNA phosphate groups. Second, this can be connected with an increase in the distance between neighboring DNA molecules in the particles of the [DNA-DAB-dendr-$(NH_2)_x$] complex. The fact that the CD band amplitude decreases with an increase in the distance between DNA molecules in the cholesteric structures formed by the phase exclusion of these molecules, despite the formation of DNA dispersion [117], favors the latter assumption.

Taking into account that the interaction between G4 molecules and DNA molecules can lead to an uncompensated positive charge in the formed dispersion particles of the [DNA-G4] complex, it can be expected that these particles will behave as independent objects. Thus, the combination of two factors—the presence of an

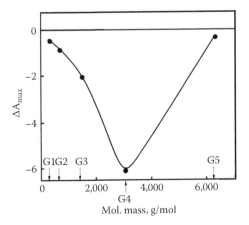

FIGURE 4.31 Dependence of the maximal amplitude of the negative band ΔA_{max} at $\lambda = 270$ nm in CD spectra of the [DNA-DAB-dendr-$(NH_2)_x$] complex dispersions on the molecular mass (generation number) of [DAB-dendr-$(NH_2)_x$].

additional positive charge in the molecules of the [DNA-G4] complex and the attraction between neighboring DNA fragments—leads to the formation of dispersion particles with an increased thermodynamic stability. If this is true, this opens the possibility for immobilizing individual CLCD particles of the [DNA-G4] complex on a nuclear membrane filter [112].

Figure 4.32 shows the dispersion particles formed from the molecules of the [DNA–G4] complex. (In Figure 4.32, C_{DNA} = 30 μg/mL; 0.6 mol/L NaCl + 0.01 mol/L phosphate buffer; pH ≈ 7.0; C_{G4} = 13.9 × 10^{-5} mol/L.) It is evident that the particles look like flattened drops with diffuse boundaries (this is "semi-rigid" particles). After the stabilization of particles with glutaraldehyde, their shape becomes close to a sphere with a diameter of 300–400 nm, as seen in Figure 4.33. (In Figure 4.33, C_{DNA} = 30 μg/mL; 0.6 mol/L NaCl + 0.01 mol/L phosphate buffer, pH ≈ 7.0; C_{G4} = 13.9 × 10^{-5} mol/L.) This result corresponds to the size determination of DNA LCD particles formed under other conditions [48] obtained by other techniques [118, 119]. Note that the borders of particles [DNA-G4] complex—treated with glutaraldehyde, which induces cross-links between available, free G4 amino groups—become more distinct, and the ratio of their height to diameter increases almost twofold (0.15–0.20 to 0.4–0.7) compared with the initial specimens. The shape of so-treated particles resembles the shape of classical rigid DNA nanoparticles. Hence, this treatment transforms the semirigid structure of particles to their rigid state. The combination of shape and nanosize of CLCD particles of the [DNA-G4] complexes and their abnormal optical activity allows one to use the term *[DNA-G4] nanoconstructions* to describe these particles.

Consequently, an electrostatic interaction between DNA molecules and [DAB-dendr-$(NH_2)_x$] molecules of various generations gives nanoconstructions of two types. The first type includes nanoconstructions with a dense packing of neighboring molecules of [DAB-dendr-$(NH_2)_x$] complexes, which display no abnormal optical activity. This suggests a hexagonal packing of these molecules in such structures. The second type includes nanoconstructions that are formed with the involvement of voluminous G4 molecules, displaying a cholesteric packing of neighboring molecules of [DNA-G4] complexes.

Hence, [DNA-polycation] nanoconstructions can be formed under certain conditions, and their properties can be regulated by various factors. Moreover, the structure of the DNA nanoconstructions and the high local concentration of DNA molecules provide the conditions for fast penetration of various compounds into the content of DNA nanoconstructions and their interaction with DNA molecules. As the result of such interactions, the high concentration of guest molecules in DNA nanoconstructions can be reached, while the DNA molecules act as a factor determining the orientation of various compounds in these nanoconstructions.

4.10   SUMMARY

The results in chapters 3 and 4 demonstrate that independently of the technologies used for design, the created liquid and rigid nanoconstructions based on linear, low-molecular-mass, rigid, double-stranded DNA (or RNA) molecules, fixed

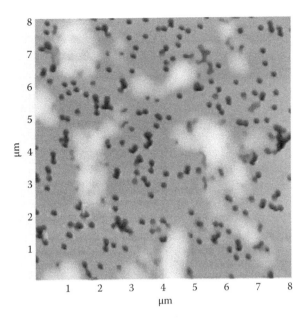

FIGURE 4.32 AFM images of the particles formed by the (DNA-G4) complex.

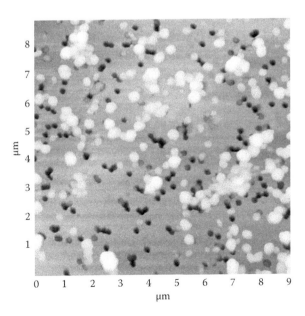

FIGURE 4.33 AFM images of the particles formed by the (DNA-G4) complex and cross-linked by glutaraldehyde.

in spatial structures of their cholesteric liquid-crystalline dispersions, display unique physicochemical properties. The combination of these properties opens a gateway for a wide application of the created NaCs in various fields of science and technology.

1. The nanoconstructions formed from DNA CLCD particles with a DNA concentration exceeding 200 mg/mL can be used as carriers for genetic material or molecules of various biologically active or chemically relevant compounds incorporated into these nanoconstructions. *Application fields: medicine and biotechnology.*
2. The nanoconstructions formed from DNA (or RNA) CLCD particles can be used as multifunctional biosensing units for optical analytical systems intended for the detection of various analytes, including nanoparticles, in various media. *Application fields: medicine, biotechnology, and ecology.*

REFERENCES

1. Leonard, M., H. Hong, N. Easwar, and H. H. Strey. 2001. Soft matter under osmotic stress. *Polymer* 42:5823–27.
2. Stanley, C. B., H. Hong, and H. H. Strey. 2005. DNA cholesteric pitch as a function of density and ionic strength. *Biophys. J.* 89:2552–57.
3. Yevdokimov, Yu. M., V. I. Salyanov, E. Gedig, and F. Spener. 1996. Formation of polymeric chelate bridges between double-stranded DNA molecules fixed in spatial structure of liquid-crystalline dispersions. *FEBS Lett.* 392:269–73.
4. Yevdokimov, Yu. M., V. I. Salyanov, F. Spener, and M. Palumbo. 1996. Adjustable cross-linking of neighbouring DNA molecules in liquid-crystalline dispersions through (daunomycin-copper) polymeric chelate complexes. *Int. J. Biol. Macromol.* 19:247–55.
5. Yevdokimov, Yu. M., V. I. Salyanov, L. V. Buligin, et al. 1997. Liquid-crystalline structure of nucleic acids: Effect of anthracycline drugs and copper ions. *J. Biomol. Struct. Dyn.* 15:97–105.
6. Yevdokimov, Yu. M., V. I. Salyanov, B. V. Mchedlishvili, et al. 2000. Double-stranded nucleic acids in liquid-crystalline dispersions as building blocks for cross-linked supramolecular structures. *Nucleosides, Nucleotides and Nucleic Acids* 19:1355–65.
7. Yevdokimov, Yu. M., S. G. Skuridin, Yu. D. Nechipurenko, et al. 2005. Nanoconstructions based on double-stranded nucleic acids. *Int. J. Biol. Macromol.* 36:103–15.
8. Wells, A. 1988. Copper, gold and silver. *Journal of Structural Chemistry* (Russian ed.) 3:224–86.
9. Kaneko, M., and E. Tsuchida. 1981. Formation, characterization and catalytic activity of polymer-metal complexes. *J. Polymer Sci. Macromol. Rev.* 16:397–81.
10. Coble, H. D., and H. F. Holtzclaw. 1974. Chelate polymers of copper (II) various dihydroxyquinoid ligands. *J. Inorg. Nucl. Chem.* 36:1049–53.
11. Yevdokimov, Yu. M., V. I. Salyanov, Yu. D. Nechipurenko, et al. 2003. Molecular constructions (superstructures) with adjustable properties based on double-stranded nucleic acids. *Molecular Biology* (Russian ed.) 37:340–55.
12. Doskočil, J., and I. Fric. 1973. Complex formation of daunomycin with double-stranded RNA. *FEBS Lett.* 37:55–58.
13. Barthelemy-Clavey V., J.-C. Maurizot, and P. J. Sicard. 1973. Etude spectrophotometrique du complexe DNA-daunorubicine. *Biochimie* 55:859–68.

14. Zakharov, M. A., L. G. Sokolovskaya, Yu. D. Nechipurenko, G. B. Lortkipanidze, and Yu. M. Yevdokimov. 2005. Formation of nanoconstructions based on double-stranded DNA. *Biophysics* (Russian ed.) 50:824–32.

15. Yevdokimov, Yu. M., S. G. Skuridin, and V. I. Salyanov. 1988. The liquid-crystalline phases of double-stranded nucleic acids in vitro and in vivo. *Liq. Crystals* 3:1443–59.

16. Wells, A. 1988. In *Structural inorganic chemistry*. Vol. 3, 223–86 (Russian ed.). Moscow: Mir.

17. Nikiforov, V. N., V. D. Kuznetsov, Yu. D. Nechipurenko, V. I. Salyanov, and Yu. M. Yevdokimov. 2005. Magnetic properties of cooper as a constituent of nanobridges formed between spatially fixed deoxyribonucleic acid molecules. *JETP Lett.* (Russian ed.) 81:327–29.

18. Yevdokimov, Yu. M., and V. V. Sytchev. 2007. Nanotechnology and nucleic acids. *Technologies of Living Systems* (Russian ed.) 4:3–27.

19. Zakharov, M. A., Yu. D. Nechipurenko, G. B. Lortkipanidze, and Yu. M. Yevdokimov. 2005. The thermodynamic stability of nanoconstructions based on the double-stranded DNA molecules. *Biophysics* (Russian ed.) 5:1036–41.

20. Grasso, D., S. Fasone, C. La Rosa, and V. Salyanov. 1991. A calorimetric study of the different thermal behaviour of DNA in the isotropic and liquid-crystalline states. *Liq. Crystals* 9:299–305.

21. van Waser, J. R., and K. Holst. 1950. Structure and properties of the condensed phosphates. I. Some general considerations about phosphoric acids. *J. Amer. Chem. Soc.* 72:639–44.

22. Strauss, U. P., D. Woodside, and P. L. Wineman. 1957. Counterion binding by polyelectrolytes. I. Exploratory electrophoresis, solubility and viscosity studies of the interaction between polyphosphates and several univalent cations. *J. Phys. Chem.* 61:1353–56.

23. Yevdokimov, Yu. M., V. I. Salyanov, and S. G. Skuridin. 2009. From liquid crystals to DNA nanoconstructions. *Molecular Biology* (Russian ed.) 43:309–26.

24. Karlik, S. J., G. L. Eichhorn, P. N. Lewis, and D. R. Crapper. 1980. Interaction of aluminium species with deoxyribonucleic acid. *Biochemistry* 19:5991–98.

25. Wedrychowski, A., W. N. Schmidt, and L. S. Hlinica. 1986. The in vivo cross-linking of proteins and DNA by heavy metals. *J. Biol. Chem.* 261:3370–76.

26. Corain, B., G. G. Bombi, and P. Zatta. 1988. Different effects of covalent compounds in aluminium toxicology. *Neurobiol. Aging* 9: 413–14.

27. Mumper, R. J., and M. Jay. 1992. Formation and stability of lanthanide complexes and their encapsulation into polymeric microspheres. *J. Phys. Chem.* 96:8626–31.

28. Qi, Y. H., Q. Y. Zhang, and L. Xu. 2002. Correlation analysis of the structure and stability constants of gadolinium(III) complexes. *J. Chem. Inf. Comput. Sci.* 42:1471–75.

29. Lessing, P. A., and A. W. Erickson. 2003. Synthesis and characterization of gadolinium phosphate neutron absorber. *J. Eur. Ceram. Soc.* 23:3049–57.

30. Zhang, P., and T. Kimura. 2006. Complexation of Eu(III) with dibutyl phosphate and tributyl phosphate. *Solvent Extract. Ion Exchange* 24:146–63.

31. Gavryushov, S. 2008. Electrostatics of B-DNA in NaCl and $CaCl_2$ solutions: Ion size, interionic correlation, and solvent dielectric saturation effects. *J. Phys. Chem. B* 112:8955–65.

32. Gersanovski, D., P. Colson, C. Houssier, and E. Fredericq. 1985. Terbium (3+) as a probe of nucleic acid structure: Does it alter the DNA conformation in solution? *Biochim. Biophys. Acta* 824:313–23.

33. Shih, J.-L. A., and R. M. Brugger. 1992. Gadolinium as a neutron capture therapy agent. *Med. Phys.* 19:733–44.

34. Martin, R. F., G. D'Cunha, M. Pardee, and B. J. Allen. 1988. Induction of double-strand breaks following neutron capture by DNA-bound 157Gd. *Int. J. Radiat. Biol.* 54:205–8.

35. Yevdokimov, Yu. M., V. I. Salyanov, O. V. Kondrashina, et al. 2005. Particles of liquid-crystalline dispersions formed by (nucleic acid-rare earth element) complexes as a potential platform for neutron capture therapy. *Int. J. Biol. Macromol.* 37:165–73.

36. Akulinichev, S. V., V. M. Skorkin, V. N. Nikiforov, et al. 2006. A new biomaterial based on the complex (DNA-Gd). 1: Determination of gadolinium concentration in the particles. *Medical Physics* (Russian ed.) 3:64–69.

37. Yevdokimov, Yu. M., V. I. Salyanov, E. V. Shtykova, et al. 2008. A transition in DNA molecule's spatial ordering due to nano-scale structural changes. *Open Nanosci. J.* 2:17–28.

38. Ha, S. C., K. Lowenhaupt, A. Rich, Y. G. Kim, and K. K. Kim. 2005. Crystal structure of a junction between B-DNA and Z-DNA reveals two extruded bases. *Nature* 437:1183–86.

39. Belyakov, V. A., V. P. Orlov, S. V. Semenov, S. G. Skuridin, and Yu. M. Yevdokimov. 1996. Comparison of calculated and observed CD spectra of liquid crystalline dispersions formed from double-stranded DNA and from DNA complexes with coloured compounds. *Liq. Crystals* 20:777–84.

40. Saeva, F. D., P. E. Sharpe, and G. R. Olin. 1973. Cholesteric liquid crystal induced circular dichroism (LCID). V. Mechanistic aspects of LCID. *J. Amer. Chem. Soc.* 95:7656–59.

41. Zipper, H., H. Brunner, J. Bernhagen, and F. Vitzthum. 2004. Investigation on DNA intercalation and surface binding by SYBR Green I, its structure determination and methodological implications. *Nucl. Acids Res.* 32:3–10.

42. Keller, D., and C. Bustamante. 1986. Theory of the interaction of light with large inhomogeneous molecular aggregates. I. Absorption. *J. Chem. Phys.* 84:2961–71.

43. Keller, D., and C. Bustamante. 1986. Theory of the interaction of light with large inhomogeneous molecular aggregates. II. Psi-type dichroism. *J. Chem. Phys.* 84:2972–80.

44. Kim, M.-H., L. Ulibarri, D. Keller, and C. Bustamante. 1986. The psi-type dichroism of large molecular aggregates. III. Calculations. *J. Chem. Phys.* 84:2981–89.

45. Belyakov, V. A., S. M. Osadchii, and V. A. Korotkov. 1986. Optics of imperfect cholesteric liquid crystals. *Crystallography Reports* (Russian ed.) 31:522–27.

46. Belyakov, V. A., V. P. Orlov, S. V. Semenov, et al. 1996. Some features of circular dichroism spectra of liquid-crystalline dispersions of double-stranded DNA molecules and DNA complexes with strained compounds. *Biophysics* (Russian ed.) 41:1044–55.

47. Li, L., J. Yang, X. Wu, C. Sun, and G. Zhou. 2003. Study of the co-luminescence effect of terbium-gadolinium-nucleic acids-cetylpyridine bromide system. *J. Lumin.* 101:141–46.

48. Yevdokimov, Yu. M., S. G. Skuridin, and G. B. Lortkipanidze. 1992. Liquid-crystalline dispersions of nucleic acids. *Liq. Crystals* 12:1–16.

49. Yevdokimov, Yu. M., V. I. Salyanov, K. A. Dembo, and F. Spener. 1999. Recognition of DNA molecules and their package in liquid crystals. *Sensory Systems* (Russian ed.) 13:158–69.

50. Grasso, D., R. G. Campisi, and C. La Rosa. 1992. Microcalorimetric measurements of thermal denaturation and renaturation processes of salmon sperm DNA in gel and liquid crystalline phases. *Thermochim. Acta* 199:239–45.

51. Besteman, K., K. Van Eijk, and S. G. Lemay. 2007. Charge inversion accompanies DNA condensation by multivalent ions. *Nature Physics* 3:641–44.

52. Rudd, L., D. J. Lee, and A. A. Kornyshev. 2007. The role of electrostatics in the B to A transition of DNA: From solution to assembly. *J. Phys.: Condens. Matter* 19:1–19.

53. Kornyshev, A. A., S. Leikin, and S. V. Malinin. 2002. Chiral electrostatic interaction and cholesteric liquid crystals of DNA. *Eur. Phys. J. E* 7:83–93.

54. Grosberg, A. Yu., T. T. Nguyen, and B. I. Shklovskii. 2002. The physics of charge inversion in chemical and biological systems. *Rev. Mod. Phys.* 74:329–45.

55. Podgornik, R., and V. A. Parsegian. 1998. Charge-fluctuation forces between rodlike polyelectrolytes: Pairwise summability reexamined. *Phys. Rev. Lett.* 80:1560–63.

56. Harris, A. B., R. D. Kamien, and T. C. Lubensky. 1999. Molecular chirality and chiral parameters. *Rev. Mod. Phys.* 71:1745–57.

57. Kornyshev, A. A., and S. Leikin. 2000. Electrostatic interaction between long, rigid helical macromolecules at all interaxial angles. *Phys. Rev. E* 62:2576–96.

58. Gelbart, W. M., R. F. Bruinsma, P. A. Pincus, and V. A. Parsegian. 2000. DNA-inspired electrostatics. *Phys. Today* 53:38–44.

59. Golo, V. L., E. I. Katz, and I. P. Kikot. 2006. Effect of dipole forces on the structure of the liquid crystalline phases of DNA. *JETP Lett.* (Russian ed.) 84:334–38.

60. Yevdokimov, Yu. M., V. I. Salyanov, S. V. Semenov, and S. G. Skuridin. 2011. *DNA Liquid-Crystalline Dispersions and Nanostructures*, ed. Yu. M. Yevdokimov. Boca Raton, FL: CRC Press.

61. Furube, A., L. Du, K. Hara, R. Katoh, and M. Tachiya. 2007. Ultrafast plasmon-induced electron transfer from gold nanorods into TiO_2 nanoparticles. *J. Amer. Chem. Soc.* 129:468–85.

62. Colvin, V. 2003. The potential environmental impact of engineered nanomaterials. *Nat. Biotechnol.* 21:1166–70.

63. Shukla, R., V. Bansal, M. Chaudhary, A. Basu, R. Bhonde, and M. Sastry. 2005. Biocompatibility of gold nanoparticles and their endocytotic fate inside the cellular compartment: A microscopic overview. *Langmuir* 21:10644–54.

64. Liu, Y., W. Meyer-Zaika, S. Franzka, G. Schmid, M. Tsoli, and H. Kuhn. 2003. Gold-cluster degradation by the transition of B-DNA into A-DNA and by formation of nanowires. *Angew. Chem. Int. Ed.* 42:2853–57.

65. Zherenkova, L. V., P. V. Komarov, and P. G. Khalatur. 2007. Modeling the process of DNA molecule fragment metallization by gold nanoparticles. *Colloid Journal* (Russian ed.) 69:753–65.

66. Herne, T. M., and M. J. Tarlov. 1997. Characterization of DNA probes immobilized on gold surface. *J. Amer. Chem. Soc.* 119:8916–20.

67. Kira, A., H. Kim, and K. Yasuda. 2009. Contribution of nanoscale curvature to number density of immobilized DNA on gold nanoparticles. *Langmuir* 25:1285–88.

68. Yevdokimov, Yu. M., Skuridin, S. G., Salyanov, V. I., Popenko, V. I., Rudoy, V. M., Dement'eva, O. V., and Shtykova, E. V. 2011. A dual effect of Au-nanoparticles on nucleic acid cholesteric liquid-crystalline particles. *J. Biomat. and Nanobiotechnol.*, 2:461–71.

69. Skuridin, S. G., V. A. Dubinskaya, E. V. Shtykova, et al. 2011. Retention of gold nanoparticles in the structure of quasinematic layers formed by DNA molecules. *Biologicheskie Membrany* (Russian ed.) 28:191–98.

70. Mirkin, C. A., R. L. Letsinger, R. C. Mucic, and J. J. Storhoff. 1996. A DNA-based method for rationally assembling nanoparticles into macroscopic materials. *Nature* 382:607–9.

71. Alivisatos, A. P., K. P. Johnsson, X. Peng, et al. 1996. Organization of "nanocrystal molecules" using DNA. *Nature*. 382:609–11.

72. Kumar, A., M. Pattarkine, M. Bhadbhade, et al. 2001. Linear superclusters of colloidal gold particles by electrostatic assembly on DNA templates. *Adv. Mater.* 13:341–44.

73. Sastrya, M., A. Kumar, S. Datar, and C. V. Dharmadhikari. 2001. DNA-mediated electrostatic assembly of gold nanoparticles into linear arrays by a simple drop-coating procedure. *Appl. Phys. Lett.* 78:2943–45.

74. Warner, M. G., and J. E. Hutchison. 2003. Linear assemblies of nanoparticles electrostatically organized on DNA scaffolds. *Nat. Mater.* 2:272–77.

75. Link, S., and M. A. El-Sayed. 1999. Spectral properties and relaxation dynamics of surface plasmon electronic oscillations in gold and silver nanodots and nanorods. *J. Phys. Chem. B* 103:8410–26.

76. Westcott, S. L., S. J. Oldenburg, T. R. Lee, and N. J. Halas. 1999. Construction of simple gold nanoparticle aggregates with controlled plasmon-plasmon interactions. *Chem. Phys. Lett.* 300:651–55.

77. Link, S., and M. A. El-Sayed. 2000. Shape and size dependence of radioactive, nonradioactive and photothermal properties of gold nanocrystals. *Int. Rev. Phys. Chem.* 19:409–53.

78. Su, K. H., Q. H. Wei, X. Zhang, J. J. Mock, D. R. Smith, and S. Schultz. 2003. Interparticle coupling effects on plasmon resonances of nanogold particles. *Nano Lett.* 3:1087–90.

79. Rechberger, W., A. Hohenau, A. Leitner, J. R. Krenn, B. Lamprecht, and F. R. Aussenegg. 2003. Optical properties of two interacting gold nanoparticles. *Opt. Commun.* 220:137–41.

80. Kamat, P. V. 2002. Photophysical, photochemical and photocatalytic aspects of metal nanoparticles. *J. Phys. Chem. B* 106:7729–44.

81. Khlebtsov, N. G., A. G. Melnikov, L. A. Dykman, and V. A. Bogatyrev. 2004. Optical properties and biomedical applications of nanostructures based on gold and silver bioconjugates. In *Photopolarimetry in Remote Sensing*, ed. G. Videen, Ya. S. Yatskiv, and M. I. Mishchenko, 265–308. NATO Science Series. II: Mathematics, physics, and chemistry, vol. 161. Dordrecht, Netherlands: Kluwer.

82. Dykman, L., V. Bogatyrev, S. Shchyogolev, and N. Khlebtsov. 2008. In *Gold nanoparticles: synthesis, properties, biomedical applications* (Russian ed.), 70–78. Moscow: Nauka.

83. Mulvaney, P. 1996. Surface plasmon spectroscopy of nanosized metal particles. *Langmuir* 12:788–800.

84. Decher, G. 1997. Fuzzy nanoassemblies: Toward layered polymeric multicomposites. *Science* 277:1232–37.

85. Yevdokimov, Yu. M. 2011. From particles of liquid-crystalline dispersions to rigid deoxyribonucleic acid nanoconstructions. *Liquid Crystals Today* 20:2–19.

86. Svergun, D. I., A. V. Semenyuk, and L. A. Feigin. 1988. Small-angle-scattering data treatment by the regularization method. *Acta Cryst.* A44 (Part 3): 244–50.

87. Rocha, A., Y. Zhou, S. Kundu, J. M. Gonzalez, S. B. Vinson, and H. Liang. 2011. In vivo observation of gold nanoparticles in the central nervous system of *Blaberus discoidalis*. *J. Nanobiotechnology* 9:5.

88. Kang, B., M. A. Mackey, and M. A. El-Sayed. 2010. Nuclear targeting of gold nanoparticles in cancer cells induces DNA damage, causing cytokinesis arrest and apoptosis. *J. Amer. Chem. Soc.* 132:1517–19.

89. Zakhidov, S. T., T. L. Marshak, E. A. Malolina, et al. 2010. Gold nanoparticles impair nuclear chromatin decondensation process in murine sperm cells in vitro. *Biologicheskie membrany* (Russian ed.) 27:349–53.

90. Boisselier, E., and D. Astruc. 2009. Gold nanoparticles in nanomedicine: Preparations, imaging, diagnostics, therapies and toxicity. *Chem. Soc. Rev.* 38:1759–82.

91. Wiwanitkit, V., A. Sereemaspun, and R. Rojanathanes. 2009. Effect of gold nanoparticles on spermatozoa: The first world report. *Fertil. Steril.* 91:e7–e8.

92. Oosawa, F. 1971. *Polyelectrolytes.* New York: Marcel Dekker.

93. Allahyarov, E., G. Gommper, and H. Lowen. 2005. DNA condensation and redissolution: Interaction between overcharged DNA molecules. *J. Phys. Condens. Matter* 17:1827–40.

94. Yoshikawa, Y., and K. Yoshikawa. 1995. Diaminoalkanes with an odd number of carbon atoms induce compaction of a single double-stranded DNA chain. *FEBS Lett.* 361:277–81.

95. Raspaud, E., M. O. de la Cruz, J.-L. Sikorav, and F. Livolant. 1998. Precipitation of DNA by polyamines: A polyelectrolyte behavior. *Biophys. J.* 74:381–93.

96. Saminathan, M., T. Antony, A. Shirahata, L. H. Sigal, T. Thomas, and T. J. Thomas. 1999. Ionic and structural specificity effect of natural and synthetic polyamines on the aggregation and resolubilization of single-, double-, and triple-stranded DNA. *Biochemistry* 38:3821–30.

97. Skuridin, S. G., V. A. Kadykov, V. S. Shashkov, Yu. M. Evdokimov, and Ya. M. Varshavsky. 1978. The formation of a compact form of DNA in solution induced by interaction with spermidine. *Molecular Biology* (Russian ed.) 12:413–20.

98. Kielland, S. L., and R. E. Williams. 1976. Water-soluble lysine-containing polypeptides. III. Sequential lysine-glycine polypeptides: A circular dichroism and electron microscopy study of annealed complexes with DNA, sonicated DNA, and denatured DNA. *Can. J. Chem.* 54:3884–94.

99. Yevdokimov, Yu. M., and V. I. Salyanov. 2003. Liquid-crystalline dispersions of complexes formed of chitosan with double-stranded nucleic acids. *Liq. Crystals* 30:1057–74.

100. Okuyama, K., K. Noguchi, T. Miyazawa, T. Yiu, and T. Ogawa. 1997. Molecular and crystal structure of hydrated chitosan. *Macromolecules* 30:5848–55.

101. Cairns, P., M. J. Miles, V. J. Morris, M. J. Ridout, G. J. Brownsey, and W. T. Winter. 1992. X-ray fibre diffraction studies of chitosan and chitosan gels. *Carbohydr. Res.* 235:23–28.

102. Hayatsu, H., T. Kubo, Y. Tanaka, and K. Negishi. 1997. Polynucleotide-chitosan complex: An insoluble but reactive form of polynucleotide. *Chem. Pharm. Bull.* 45:1363–68.

103. Lee, S.-T., F.-L. Mi, Y.-J. Shen, and S.-S Shyu. 2001. Equilibrium and kinetic studies of copper (II) ion uptake by chitosan-tripolyphosphate chelating resin. *Polymer* 42:1879–92.

104. Domard, A. 1987. pH and CD measurements on a fully deacetylated chitosan: Application to copper (II) polymer interactions. *Int. J. Biol. Macromol.* 9:98–104.

105. Monteiro, O. A. C., and C. Arnoldi. 1999. Some thermodynamic data on copper-chitin and copper-chitosan biopolymer interactions. *J. Coll. Interface Sci.* 212:212–219.

106. Inoue, K., Y. Baba, and K. Yoshizuka. 1993. Adsorption of metal ions on chitosan and cross-linked copper (II)–complexed chitosan. *Bull. Chem. Soc. Jpn.* 66:2915–21.

107. Lima, I. S., and C. A. Airoldi. 2004. A thermodynamic investigation on chitosan-divalent cation interaction. *Thermochim. Acta* 421:125–31.

108. Rhazi, M., J. Derbrieres, E. Talaimate, et al. 2002. Contribution to the study of the complexation of copper by chitosan and oligomers. *Polymer* 43:1267–76.

109. Govorun, E. N., K. B. Zeldovich, and A. R. Khokhlov. 2003. Structure of charged poly(propylene imine) dendrimers: Theoretical investigation. *Macromol. Theory Simul.* 12:705–13.

110. Fant, K., E. K. Esbjörner, P. Lincoln, and B. Norden. 2008. DNA condensation by PAMAM dendrimers: Self-assembly characteristics and effect on transcription. *Biochemistry* 47:1732–40.

111. Popenko, V. I., S. G. Skuridin, V. I. Salyanov, et al. 2007. "Rigid" cholesterics formed by complexes of double-stranded DNA with dendrimers. Paper presented at the 15th Russian symposium on scanning electron microscopy and analytical methods for the study of solids (Russian ed.), 3017. Chernogolovka, Russia.

112. Skuridin, S. G., V. I. Popenko, V. A. Dubinskaya, V. A. Bykov, and Yu. M. Yevdokimov. 2009. Liquid-crystalline dispersions formed by complexes of linear double-stranded DNA molecules with dendrimers. *Molecular Biology* (Russian ed.) 43:492–504.

113. Kabanov, V. A., V. G. Sergeyev, O. A. Pyshkina, et al. 2000. Interpolyelectrolyte complexes formed by DNA and astramol poly(propylene imine) dendrimers. *Macromolecules* 33:9587–93.

114. Terentiev, E. M. 1995. Stability of liquid-crystalline macroemulsions. *Europhys. Lett.* 32:604–12.

115. de la Cruz, M. O., L. Belloni, M. Delsanti, J. P. Dalbiez, O. Spalla, and M. Drifford. 1995. Precipitation of highly charged polyelectrolyte solutions in the presence of multivalent salts. *J. Chem. Phys.* 103:5781–91.

116. Stilck, J. F., Y. Levin, and J. J. Arenzon. 2002. Thermodynamic properties of a simple model of like-charged attracting rods. *J. Stat. Phys.* 106:287–99.

117. Yevdokimov, Yu. M., S. G. Skuridin, S. V. Semenov, V. I. Salyanov, and G. B. Lortkipanidze. 1998. Stabilization of the optical properties of cholesteric liquid-crystalline dispersions. *Biophysics* (Russian ed.) 43:240–52.

118. Skuridin, S. G., E. V. Shtykova, and Yu. M. Evdokimov. 1984. Formation kinetics of optically active liquid-crystalline microphases of low-molecular-weight DNA. *Biophysics* (Russian ed.) 28:337–38.

119. Yevdokimov, Yu. M., S. G. Skuridin, and N. M. Akimenko. 1984. Liquid-crystalline microphases of low-molecular-weight double-stranded nucleic acids and synthetic polynucleotides. *Polymer Science, Series A* (Russian ed.) 26:2403–10.

5 Viral Particles as "Rigid" Biological Nanoconstructions: Their Creation and Medical Application Prospects

D. Yu. Logunov, B. S. Naroditsky, and A. L. Gintsburg

In the previous chapters of this book, DNA molecules were examined as biopolymers, whose physicochemical properties can be used to create various nanostructures and nanoconstructions that can be controlled under laboratory conditions. The sizes and characteristics of the formed structures correspond to an idea about specific properties of nanoparticles.

However, the uniqueness of double-stranded DNA as a nanotechnology object consists, perhaps, in another property of this biopolymer. *DNA is the most widely spread natural information carrier* that stores coded data on the primary structure of proteins. The latter are well known and are capable of forming various multimodule complexes, the size of which may be on the order of several hundred nanometers. Thus, it is possible, at least in theory, to use both physicochemical and informational peculiarities of DNA for encoding and targeted formation of protein-based structures. Knowing that nature has a much longer experience in dealing with nanotechnologies, one might realize that there is a great number of biogenic nanoobjects, based on DNA enclosed in various types of containers (consisting of proteins, lipids, and carbohydrates in different combinations). At the same time, the DNA molecule determines the structure and the composition of such containers to a significant extent. The relatively simple forms of biological nanosized objects are bacteriophages and viruses. Hence in this chapter we attempt to show how these objects can be generated under conditions existing in living cells.

5.1 A "PEG-LIKE SITUATION" IN BIOLOGICAL SYSTEMS AND PECULIARITIES OF BIOLOGICAL NANOCONSTRUCTIONS

While studying the assembly of bacteriophage T4 in *E. coli* cells, Laemmli, Paulson, and Hitchins [1] focused their attention on the process of condensing phage double-stranded DNA under cell conditions. It was shown that at a definite stage of the cell cycle, which is of interest from a physicochemical point of view, splitting of protein P22 takes place. This is cleaved into two low-molecular-mass proteins, II and VII. These proteins contain glutamine and asparagines amino-acid residues at a level of 80% and 40%, respectively. Within the cell, these proteins are negatively charged and do not form stoichiometric complexes with DNA. The concentration of proteins II and VII in a cell at the packing site of the phage particles is estimated to reach about 500 mg/mL, which, in combination with the other properties, induces the phase exclusion of DNA molecules and the formation of a DNA phase consisting of one molecule. This implies that under cell conditions, the water-soluble biopolymers (proteins) create conditions that correspond to those when nucleic acid molecules are excluded from water–salt–PEG-containing solutions (PEG is polyethylene glycol); such a situation was termed a *PEG-like situation.*

Taking into account the high concentration of glutamine and asparagines amino acid residues in proteins II and VII, it was shown [1] that addition of poly(glutamine) and poly(asparagine) acids (as well as the PEG addition) to the water–salt solution containing double-stranded DNA molecules results following a change in its hydrodynamic properties, which suggests the formation of the DNA phase. Immediately after exclusion of the DNA molecules (or coordinated with this process), the protein capsule (phage head) is formed around the condensed DNA molecule from appropriate proteins. Additional processes result in formation of a native phage particle. As a result of this procedure, the solubility of the DNA molecule is decreased and, finally, the rigid bacteriophage particle appears. (One can suppose that a similar mechanism is at work in the process of creating other rigid biological nanosized objects, for instance, viruses.)

Therefore, the process of formation of the bacteriophage particles in vivo appears to occur in a similar fashion to that of the liquid or even rigid DNA liquid-crystalline nanoconstructions formed in vitro and described in the previous chapters of this book. This means that results describing the formation of liquid-crystalline dispersions (LCDs) of nucleic acids allow one to model and conserve the phase of the nucleic acids in biological systems in vitro and to elucidate the factors for direct control of the packing process of the genetic material in vivo.

From a physicochemical point of view, a few specific features of biological objects (for instance, bacteriophages, viruses, chromosomes) containing double-stranded nucleic acids are of interest:

1. High local concentration of nucleic acids (from 200 to 800 mg/mL [2]
2. The local ordering of a nucleic acid molecule (or its segments) [2]
3. The specific mode of the organization of molecules of nucleic acids [2]
4. The reconstruction of the model of DNA packing, which led to the conclusion that the DNA molecules in protozoa chromosomes are packed similarly to cholesteric liquid crystals [3]

5. The recent investigation of DNA molecules packing in the capsids of bacteriophages λ, SPP1, and T5 [4], which demonstrated not only the presence of hexagonal (or quasi hexagonal) crystalline domains in the packed DNA molecules, but phase transition between various crystalline DNA forms [5]

The biogenic nanosized rigid objects mentioned here are viruses, which constitute a specific form of life, and their individual particles are usually referred to as virions. The size of virions varies, as a rule, from 20 to 100 nm (but for certain viruses, the size of the capsid exceeds 100 nm and may be as large as 700 nm).The protein capsule (capsid) of each virion is the genomic DNA of the virus, which is responsible for reproducing identical copies of the virion during the process of a viral infection. The reproduction rate may exceed 1,000 virions per infected cell. The reproduction of the virus consists of two simultaneous processes: the replication of the viral DNA (multiplication of copies of the viral genome) and the biosynthesis of regulatory and structural proteins encrypted in it. The viral structural proteins have an important property: the ability to self-assemble, due to bottom-up technology, into homo- or heterogeneous nanostructures of a high order (capsids), which is necessary for immobilization of the DNA molecule. After immobilization of the DNA molecules inside the viral capsid, the viral particle (virion) exists as a rigid biological object. (Taking into account the size of the viral particles (about 100 nm), one can say that these objects can be considered as a rigid DNA-based nanoconstruction.) The characteristic structural peculiarities of the virion capsid determine the subsequent ability of the whole viral particle to penetrate the cells and organs of the host organism, which are specific for each viral species.

The rigid viral particles exhibit a number of properties that favor their use as a convenient technological nanoplatform (based on DNA):

- A relative simplicity of the structural organization
- A size and shape uniformity of particles within a viral species
- A high degree of symmetry and polyvalency of the particles
- An ability to control the size, shape, and properties of viral structures at the genome level (using genetic engineering techniques)
- A retention of a stable structure for a long time
- A natural mechanism of packing of genetic information carrier into a protein shell (capsid)
- A natural mechanism of cell penetration

Notably, viruses as nanostructures have some additional unique features that are of key importance for solving a number of problems under in vivo conditions. Viruses are partially resistant to aggressive media of the living organism; they are spread over its various compartments by the blood flow; they target (depending on the tropism) certain types of organs and tissues; and they are both biocompatible (replication-deficient virions are safely administered to humans and animals within a broad range of doses—up to 10^{13} virions per dose) and biodegradable (the washout is completed within three–seven days via the activity of cells eliminating the virions).

In other words, the properties of viral particles favor their use as dedicated nano-structures or nanoconstructions not only under in vitro conditions (this problem was considered in previous chapters), but in various in vivo systems (tissues, organs, and organisms), where the existence and functioning of such structures are determined by a number of additional requirements. All of these properties make viruses one of the most probable candidates to become the background, based on which nanoconstructions for medical use (development of drugs for gene therapy, vaccines, and new technologies for noninvasive diagnostics of diseases) will be created in the nearest future.

The chemical structure of viral particles contains a large amount of functional groups that can be used for formation of complex structures. Moreover, these groups favor their modification by both physicochemical methods and genetic engineering techniques. At the realization of the first method, various components (metal nanoparticles, polymer molecules, etc.) are incorporated into the structure of natural viral particles. Such modified viral particles have already shown significant promise in experiments on diagnostics and therapy of a number of diseases (including neoplastic diseases). The disadvantage of this method is that in order to obtain modified viral particles, it is necessary to conjugate the viral particles with polymers or particles each time prior to use of the preparation.

The genetic-engineering approach makes it possible to obtain various viral particles by changing the structure of the viral genome. Such changes confer new features (e.g., modified tissue specificity and specific therapeutic properties) upon the viral nanoparticles and, at the same time, retain self-replication ability in specialized cells. Genetically modified viral nanoparticles can be produced in preparative amounts by specialized cell cultures. (Each particle thus produced carries by default all changes without the necessity of additional manipulations.)

Currently, viral particles for medical application with various parameters are based on recombinant adenoviruses [6–8], retroviruses [7–9], adeno-associated viruses [7, 8], herpes virus [7, 8], and certain other species [7, 8]. In this chapter, the principal characteristics, methods of preparation, and prospects of medical use of adenovirus-based nanoparticles (recombinant adenoviruses) will be examined to illustrate how, under cell conditions, by means of genetic engineering, biological factors can influence the formation of biologically active or biologically inactive rigid DNA-based nanoconstructions.

5.2 METHODS FOR OBTAINING RECOMBINANT ADENOVIRUS "RIGID" NANOSTRUCTURES

An adenovirus virion consists of a double-stranded DNA molecule enclosed in a protein shell or capsid (Figure 5.1). One can add here that, unfortunately, the structural characteristics of the double-stranded DNA molecule and the mode of its packing inside the adenovirus capsid are practically unknown. The virions are isometric icosahedron-shaped particles ranging in size from 70 to 90 nm [10].

The molecular mass of a virion is $(170–175) \times 10^6$ g/mol; its constant sedimentation is 560 S. The structure of adenoviral virions is retained at high ionic strength, acidic pH (as low as pH 4), and at temperatures in the range from 10°C to 85°C.

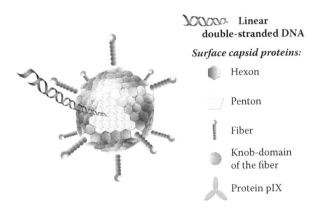

Linear
double-stranded DNA

Surface capsid proteins:

Hexon

Penton

Fiber

Knob-domain
of the fiber

Protein pIX

FIGURE 5.1 (See color insert) Structure of adenovirus virion.

A single adenoviral virion contains about 2,700 polypeptide molecules of thirteen types [11]. The molecule of the viral DNA and six types of proteins (pV, pVII, Mu, pIVa2, terminal protein, and 23K protease) form the inner deoxyribonucleoprotein (DNP) core of the virion. The other seven types of proteins form the viral capsid (Figure 5.1). The principal components of the capsid are hexones (240 capsomeres). The vertex capsomeres (penton) carry one or two filamentous spikes (or fibers) of 10 to 37 nm in length. The viral capsid also contains several minor proteins: pIIIa, pVI, pVIII, and pIX.

Adenoviruses have also been characterized in sufficient detail at the genetic level (Figure 5.2). The genomes of most adenoviruses have been completely sequenced. The DNA replication mechanisms and gene expression have been determined as well as the function of the majority of adenoviral proteins. The main mechanisms of interactions of adenoviruses with host organisms have also been studied at the molecular and cellular levels [12].

The detailed data on the structure of adenoviruses and their physicochemical and biological properties make it possible to manipulate both whole virions and their parts: individual genes and domains of capsid proteins. Such manipulations provide a way of creating recombinant adenoviral nanoconstructions (RAVNs) with new pre-programmed properties. Processing from the structure of an adenoviral virion, a rigid

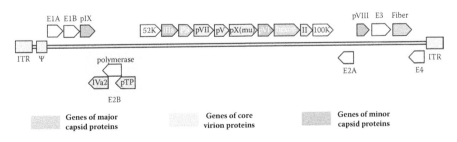

FIGURE 5.2 (See color insert) Schematic representation of adenovirus genome. Legend keys: ITR, inverted terminal repeat; P, protease; Ψ, packing signal; PB, penton base.

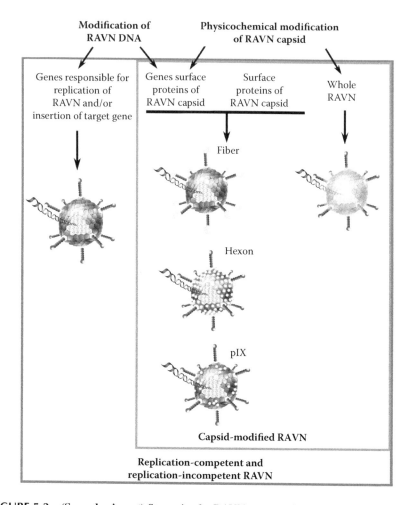

FIGURE 5.3   (**See color insert**) Strategies for RAVN construction.

RAVN can be considered as a container consisting of adenovirus capsid proteins, which serves as a vector to transfer the capsid-enclosed genetic information into cells.

The two major approaches to modifying adenoviral virions (Figure 5.3) for the purpose of obtaining RAVNs consist of genetic manipulations (of the viral DNA) and physicochemical treatment (of proteins, their groups, or the whole capsid).

5.3   RECOMBINANT ADENOVIRUS NANOSTRUCTURES WITH A MODIFIED GENOME

The RAVNs with modified genomes are obtained by targeted modification of the DNA of the initial adenovirus. Methods of genetic engineering are used to include foreign nucleotide sequences into the adenoviral DNA, change its nucleotide composition, or delete some of the original elements. Adenoviral DNA contains genes responsible for the viral proliferation (genome replication, expression of viral genes, and protection against antiviral activity of the host organism) and genes encoding structural proteins of the capsid. Changes made to the genes encoding capsid proteins produce RAVNs with targeted modifications of the structure and properties of the nanocontainer. RAVNs with modified capsid proteins are referred to as *capsid-modified* (Figure 5.3). The selection of the foreign genetic material (incorporated into adenoviral DNA for the purpose of its modification), which will be delivered to the cells by the nanocontainer, is governed by the intended use of the RAVN. At the same time, for safety reasons, genes responsible for the replication of the virus are deleted from the adenoviral DNA. As a result, replication-incompetent RAVNs (Figure 5.3) are formed, which deliver foreign genetic information to the target cells and enable them to express the relevant genes. In some cases, the genes responsible for the genome replication are intentionally preserved. The regulatory elements controlling their expression are substituted for by other elements, the function of which depends on the presence of specific "inductor" molecules in the target cells. As a result, replication-competent RAVNs (Figure 5.3) capable of replicating under strictly defined conditions are obtained.

5.4 REPLICATION-COMPETENT RECOMBINANT ADENOVIRUS NANOSTRUCTURES

The life cycle of viruses is confined to the intracellular space. Upon entering the cells, adenoviruses induce an active infection, which eventually causes the infected cell to undergo lysis, resulting in the release of the viral progeny. The infection is underlain by proliferation of the virus within the cell, which consists of replicating (copying) the viral genome followed by packing of each copy into the protein capsid. The ability of adenoviruses to kill the cells they infect was used in the creation of RAVNs targeted at the destruction of tumor cells. The capacity for proliferation, restricted solely to tumor cells, is the major characteristic of such RAVNs (Figure 5.4).

For instance, RAVNs have been created to replicate specifically in prostate gland tumor cells and destroy them. To obtain such RAVNs, the conditions of one of the genes necessary for the replication of the genome (E1A gene) have been changed. The natural promoter of this gene was replaced in the RAVN genome by prostate-specific antigen (PSA), the concentration of which increases considerably in prostate cancer cells. Proliferation of the resulting RAVNs depended entirely on the presence within the target cells of factors increasing PSA expression. The RAVNs thus obtained were capable of infecting only prostate cancer cells. Those RAVNs that entered healthy cells underwent rapid elimination from the organism [13].

Yet another type of RAVN was created to destroy the cells of tumors arising from p53 gene malfunction. This gene is responsible for self-destruction of cells with functional abnormalities, including those caused by viral infection. The penetration of a virus into a normal cell increases the p53 expression level, which in turn causes the cell to undergo apoptosis, resulting in elimination of the infection agent. One of the adenoviral proteins (E1B gene product with a molecular mass of 55×10^3 g/mol) is capable of binding p53 gene promoter and thereby inhibiting its expression.

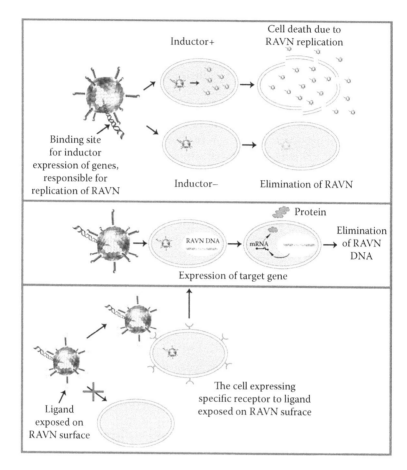

FIGURE 5.4   (See color insert) Effects of RAVNs with modified genomes: (A) replication-competent; (B) replication-incompetent; (C) capsid-modified.

Deletion of E1B from the genome produces RAVNs that cannot induce infection in cells with functionally active p53. Conversely, if p53 within a cell is not functioning (and, therefore, its tumor-suppression potential is severely reduced), the RAVN in question will successfully replicate in such a cell, cause its destruction, and release newly formed RAVNs capable of killing tumor cells [14].

5.5   REPLICATION-INCOMPETENT RECOMBINANT ADENOVIRUS NANOSTRUCTURES

As mentioned previously, an adenoviral infection is based on the replication of the viral genome followed by the expression of all of the genes encrypted in it. The ability of adenoviruses to express all genes constituting the viral genome has become the base for construction of RAVNs intended for transferring foreign genetic material to certain target cells, with subsequent expression. RAVNs created for this

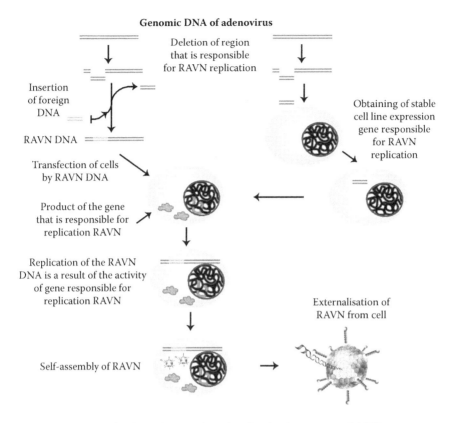

Genomic DNA of adenovirus

FIGURE 5.5   **(See color insert)** Formation of replication-incompetent RAVNs.

purpose should be safe and unable to cause the development of an infection process. Replication-incompetent RAVNs, obtained by deleting the genes responsible for its replication from the original adenoviral genome, meet this criterion (Figure 5.4). It should be noted that, without deleting any fragments from the adenovirus, only small amounts of DNA (5% of the genome size) can be incorporated into it additionally. Upon increasing the volume of foreign genetic information intended for transfer into target cells using RAVNs, larger segments of adenoviral DNA must be deleted to accommodate the incorporated sequences.

For instance, replication-incompetent RAVNs can be obtained by deleting the genes of E1 area of adenoviral genome (Figure 5.5). Under in vivo conditions, the RAVNs thus obtained are unable to replicate in the cells of the natural host in the absence of the products of the deleted E1 area. Cell lines such as 293, 911, and PerC6 for obtaining and accumulation such RAVNs have been created by introducing into their genomic DNAs the region deleted from the genomic DNA of the adenovirus. These cells efficiently express the incorporated adenoviral genes, thereby compensating for their absence in RAVN DNA (Figure 5.5). The size of the inserts in RAVNs is also increased by deletions in the area E3, responsible for the protection of adenovirus-infected cells from the antiviral activity of the immune system of the organism [15].

Human adenovirus serotype 5 served as a base for obtaining RAVNs that lack all viral genes and consist of two inverted terminal repeats (the left one and the right one) and the signal for DNA packing into the capsid. Theoretically, such RAVNs may contain up to 35,000 base pairs of foreign DNA [16]. The accumulation of these minimum vectors is achieved using an adenovirus helper as a complement providing *in trans* all viral functions necessary for replication and virion assembly. However, these RAVNs are not widely used yet, as the preparations contain admixtures of a "helper."

5.6 CAPSID-MODIFIED RECOMBINANT ADENOVIRUS NANOSTRUCTURES

Adenoviruses are capable of entering a wide range of cells, including endotheliocytes, myocytes, respiratory epithelium, primary tumor cells, hematopoietic cells, dendrite cells, and other types of body cells [17–26]. Different adenovirus serotypes differ in their ability to enter particular cell types. Thus, adenovirus serotypes 2, 4, 5, and 7 exhibit maximum affinities for lung epithelium and other cells of the respiratory tract, as opposed to serotypes 40 and 41, which preferably infect cells of the gastrointestinal tract. The adenovirus interaction with (and subsequent entry into) particular cells of the organism are determined by the structure of surface proteins of capsid. The main participants in these processes are major capsid components (hexon, penton base, and fiber proteins). Obtaining target-specific RAVNs is the main purpose of modifying the structure of adenovirus capsid proteins. Target-specific RAVNs, as their name suggests, are capable of targeting and entering specific cells of the organism (Figure 5.4).

Currently, there are several strategies for creating target-specific RAVNs. The first one is the change in the structure of fiber. Fiber plays a crucial role in determining virus specificity for a particular cell type, because it is responsible for binding of the virion to the surface of the cell. Fiber proteins of all adenoviruses are highly homologous to each other over the whole sequence, with the exception of the terminal domain. This structural feature creates an opportunity for constructing chimeric RAVNs, in which fiber as a whole (or its terminal domain) is replaced by the homologue of a different adenovirus serotype. Using this approach, RAVNs were obtained that exhibit specificity for CD34-positive cells (substitution of serotype 35 fiber for serotype 5 fiber), smooth muscle cells (substitution of serotype 16 fiber for serotype 5 fiber), and epithelial cells of upper respiratory passages (substitution of serotype 17 fiber for serotype 2 fiber). A chimeric RAVN obtained by replacing the terminal domain of serotype 5 fiber with a homologous sequence of serotype 3 fiber acquired increased affinity for ovarian cancer cells and squamous carcinoma cells [27].

The second strategy for constructing target-specific RAVNs involves incorporation into fiber structure of ligands (mostly peptides) for certain cell surface receptors. Introduction of the integrin-binding Arg-Gly-Asp (RGD) motif or a lysine heptamer into the terminal portion of fiber produced RAVNs that efficiently interacted with cells not recognized by the original (unmodified) precursor RAVNs. For example, serotype 5–based RAVNs containing the RGD motif were capable of entering ovarian cancer cells, pancreatic carcinoma cells, and certain cells of head and neck cancer, whereas the presence in fiber sequences of the lysine heptamer enabled entry into macrophages, smooth muscle cells, endo-theliocytes, T-cells, glioma cells, and myeloma cells [27]. A similar result was obtained by introducing the RGD motif into fiber of avian adenovirus serotype 1 [28]. The terminal portion of fiber was also modified by bacteriophage peptides (e.g., the peptide SIGYLPLP), and the resulting RAVNs exhibited increased specificity for blood vessel endothelial cells, transferring receptors, and smooth muscle cells [27].

The diversity of adenoviruses and viral peptide ligands is not sufficient, however, for obtaining a broad panel of the necessary target-specific RAVNs. To solve this problem, a third strategy has been developed, within which fiber is replaced com-pletely by structural components of other viruses or bacteria as well as by ligand mol-ecules. Thus, replacing fiber by bacteriophage T4 fibritin, which contains a histidine hexamer, favors RAVN penetration into cells expressing histidine hexamer receptors. Replacement of fiber by fibritin carrying a CD40 ligand insertion was reported to produce RAVNs, which interacted with CD40-expressing dendrite cells [29].

Capsid proteins modified for the purpose of obtaining target-specific RAVNs include, in addition to fiber, hexon, penton base, and pIX. The adenoviral capsid protein pIX stabilizes hexon–hexon interactions, thereby supporting virion integrity. Because the C-terminal region of pIX is located at the capsid surface, target-specific RAVNs are largely obtained by introducing ligand insertions into this particular portion of the protein. The ligands used for modifying the C-terminal portion of pIX included the RGD-motif and the lysine heptamer peptide. The resulting RAVNs were capable of efficiently entering into cells that express heparan sulfate on their surface [29].

5.7 PHYSICOCHEMICAL MODIFICATIONS OF RECOMBINANT ADENOVIRUS PARTICLES

Administration of RAVNs is frequently associated with their rapid elimination from the organism, which is effected by liver cells (Kupffer cells), macrophages, and monocytes [30]. The effect is most pronounced when RAVNs are administered systemically. In this respect, RAVNs are often administered in excessive amounts, which may lead to a number of negative results, such as retention of the majority of the particles in hepatocytes [31]. The immune response elicited by capsid proteins is yet another negative consequence of RAVN administration [32].

Several approaches have been developed to eliminate the negative effects and diminish the dose of the introduced RAVN. One of them involves using geneti-cally modified target-specific RAVNs. Another approach is based on the creation of RAVNs with physicochemical modifications of capsid proteins.

RAVNs modified with polymers exhibit maximum efficiency and safety. Examples of such modifying agents include N-(2-hydroxypropyl)-methacrylamide (HPMA) and PEG [33]. HMPA-based polymers have been used for more than a decade as nonimmunogenic and nontoxic vehicles for delivery of chemotherapy agents [34]. Polymer modification attenuates the immune response elicited by coat proteins of

RAVNs; it is believed that this effect is underlain by suppression of their interactions with cells of the immune system. However, the efficiency of RAVN interactions with other cells is also decreased. In order to preserve the capacity for target-cell interactions in polymer-modified RAVNs, various ligands are introduced into the modifying polymer, such as peptides, proteins, or antibodies [35–37]. For example, the integrin-binding RGD motif was covalently introduced into PEG-modified RAVNs. In vivo experiments demonstrated that the interaction with target cells is more efficient in RAVNs obtained by double modification (with PEG and the RGD motifs) than in genetically modified RAVNs with the RGD motif in the sequence of fiber [37].

Polymer particles interact with RAVN surfaces at lysine residues, which are present in all major capsid proteins. It is also possible to use the same amino acid residues to modify RAVNs with other molecules, e.g., peptides, carbohydrates, biotin, and certain fluorophores.

The application of combined strategies of genetic and physicochemical modification makes it possible to create RAVNs that carry metal nanoparticles at the surface of the protein coat. For instance, genetically modified RAVNs with histidine hexamer inserts in hexon or pXI sequences may efficiently bind gold nanoparticles. Reactive nickel nitrile acetate groups introduced artificially into gold nanoparticles are capable of interacting efficiently with histidine surface residues of RAVNs [38]. The resulting RAVNs may be used in oncology patients for delivering gold particles to tumor cells, because gold is known to potentiate the effects of radiotherapy.

Therefore, rigid RAVNs are a promising platform for the creation a broad range of medications with diverse applications. The properties of these nanoconstructions can be changed in accordance with the aim of the researcher by manipulating the molecules of the initial adenovirus as well as by incorporating other biogenic and abiogenic components into the RAVNs.

5.8 APPLICATION OF RECOMBINANT ADENOVIRUS NANOSTRUCTURES IN MEDICINE

5.8.1 USE OF RAVNs IN GENE THERAPY

Contemporary methods for treatment of various diseases of humans and animals are connected with the administration of deficient metabolic products into the recipient body. For instance, a common practice is to introduce an expression product of a deficient gene (e.g., an enzyme or a peptide hormone) into the organism. Novel treatment strategies are instead based on introducing a functionally active gene or a number of genes that control the synthesis of these proteins (e.g., by encoding the relevant enzymes). RAVNs have been used over decades for delivery of such genes to the organism. RAVN-based therapeutic agents are developed by introducing an expression cassette that carries the target gene (i.e., the gene of therapeutic intervention) into the adenoviral DNA.

The following key factors determine the efficiency of a RAVN preparation:

• The level of expression of the target gene

- The duration of circulation of the product of the target gene in the organism (i.e., the period during which the product remains in the body)
- The RAVN specificity for those cells, tissues, and organs, the entry into which ensures attaining the therapeutic effect in the best possible way
- The development of drugs for gene therapy that is largely based on the use of replication-deficient target-specific RAVNs

Increased expression of the target gene in a RAVN is usually achieved by selecting an optimum set of regulatory elements for inclusion into the expression cassette (which contains, in addition to the target gene, regulatory elements). The promoter is a major regulatory element affecting the expression level. It is the promoter that determines whether the gene will be expressed in a defined cell population or in the majority of cell types of the organism. The promoter currently used most widely, obtained from the genome of the human cytomegaly virus, is capable of enabling high levels of gene expression in almost any cell type. Increased levels of gene expression are achieved by including some additional 5'- or 3'-nontranslatable regions of regulatory elements into the RAVN expression cassette, such as the polypurine (A)-rich sequence, the posttranscriptional regulatory element of woodchuck hepatitis virus, or the internal ribosome-binding site [39]. The late transcription region of adenovirus genome contains its own 5'-nontranslatable element, which is termed *tripartite leader* (TPL) and *bipartite leader* (BPL) in *Mastadenoviridae* and *Aviadenoviridae*, respectively [40, 41]. The use of TPL and BPL as supplementary regulatory elements of expression cassettes within RAVNs produces a two- to threefold increase in the expression level of the transgene [42].

The physicochemical modifications of RAVN capsid, described previously, which attenuate RAVN interactions with cells of the immune system, ensure prolonged circulation in the organism of the target gene product. It should be noted, however, that residual expression of adenoviral genes present in the DNA of RAVNs may also exert cytotoxic effects and elicit host immune responses [43, 44], resulting in decreased duration of the expression of the delivered transgene. These side effects may be overcome by deleting certain regions of the viral genome. For example, transient transgene expression may be prolonged by deleting regions responsible for adenoviral genome replication and gene expression (E1 and E2), which prevent the synthesis of RAVN capsid proteins.

Treatment of diseases caused by genetic defects requires maximum duration of the expression of the gene of therapeutic intervention (e.g., over a period of several years). Lack of integration into the cell genome of adenoviral DNA causes it (and the transgene as its part) to undergo gradual (over a period of several weeks) elimination from the organism. This problem can be addressed by ensuring incorporation of RAVN DNA into the genome of host cells. For this purpose, hybrid recombinant nanoparticles were developed based on the use of adenovirus and adeno-associated virus. Adeno-associated virus is integrated into the genome of the cells, and this prolongs the expression of the therapeutic gene harbored in the viral DNA.

Processing from above, it can be assumed that most areas of RAVN-based gene therapy can contain the following set of structural elements:

- DNA carrying the expression cassette with a set of regulatory elements providing the sufficient level of the target gene expression needed to exert the
  therapeutic effect
- A capsule (of viral or nonviral origin) protecting the RAVN from the effects
  of the immune cells and providing the target gene delivery

Attempts to construct such RAVNs are currently under way in many research
institutions throughout the world. The spectrum of diseases targeted by the RAVNs
under development is extremely broad. It includes both genetic diseases and conditions acquired throughout life, including infections [45, 46]. For instance, human
adenovirus serotype 5–based replication-incompetent RAVNs were constructed,
the DNA of which carries an expression cassette harboring the human lactoferrin
gene. Lactoferrin exerts antibacterial immunomodulating effects and also exhibits
anti-inflammatory, detoxifying, and antioxidant activities. Administration of these
RAVNs demonstrated their ability to enter diverse cell types and efficiently express
the target gene [47].

RAVNs created for application in gene therapy have good prospects for clinical applications. Currently, two RAVN-based drugs are already approved for clinical use in China as antineoplastic agents: Gendicine (Sibiono GeneTech Co.) and
H101 (Sunway Biotech Co.) [48, 49]. The efficiency of two other RAVN-based drugs
with similar effects have been studied in the United States: Advexin® (Introgen
Therapeutics) and ONYX-015 (Onyx Pharmaceuticals) are currently under phase II
and phase I clinical trials. Advexin is a RAVN, the DNA of which incorporates an
expression cassette with a standard set of regulatory elements (human cytomegaly
virus early transcription promoter and SV-40 polyadenylation signal) and the therapeutic gene encoding the p53 tumor suppressor protein. This protein responds to
diverse disturbances in the genome of a cell by triggering a cascade of reactions
leading to its death. The protein also acts to suppress the formation of malignant
tumors. Mutations in the gene encoding p53 are found in approximately 50% of cancer cells. RAVN-assisted introduction of a mutation-free copy of p53 gene into tumor
cells serves as a therapeutic intervention. Advexin has successfully completed phase
I clinical trials in patients with five tumor types, and a phase III trial is currently
under way in a cohort of head and neck cancer patients (Table 5.1).

Two years of clinical trials of RAVN-expressed genes encoding endothelial
growth factor and angiogenin in patients suffering amyotrophic lateral sclerosis
demonstrated adequate tolerability, safety, and efficiency of these preparations (The
neurotrophic effects exerted by the expressed genes were shown to prolong the life
of the patients.) [50]. RAVNs carrying the gene of angiogenin proved to be efficient
as an agent for experimental treatment of chronic lower-limb ischemia (in rats and
patient volunteers); the decrease in the severity of the ischemia was demonstrated by
instrumental examination [51, 52].

5.8.2 Use of RAVNs in Vaccination

RAVNs expressing antigens of a pathogen are capable of eliciting in the organism
cellular (Th1-dependent) and humoral (Th2-dependent) immune responses to the

TABLE 5.1

Clinical Trials of Advexin as an Agent for Treatment of Various Neoplastic Diseases

Type of Tumor	Phase of Clinical Trials
Head and neck cancer	III
Non-small cell lung cancer	II
Breast cancer	II
Esophageal cancer	II
Prostate cancer	I
Testicular cancer	I
Bladder cancer	I
Brain cancer	I

pathogen. This property of RAVNs is a crucial factor in the development of vaccines against intracellular pathogens. Development of RAVN-based vaccines against a number of bacterial (tularemia, tuberculosis, and brucellosis) and viral (influenza, HIV infection, papillomatosis, rabies, Ebola hemorrhagic fever, etc.) diseases is currently underway.

In particular, RAVNs proved to be efficient in the development of candidate vaccines against diverse serotypes of influenza virus (including the avian influenza virus H5N1). As of today, a RAVN-based vaccine against influenza A has successfully completed phase I clinical trials in the United States. The study involved twenty-four volunteers and was administered with RAVN preparations intranasally or subcutaneously. RAVNs carrying the gene of influenza virus hemagglutinin were shown to be safe for humans and highly immunogenic, suggesting further use of the preparation as a vaccine [53]. Experiments in which the antigenic determinant of influenza virus hemagglutinin was incorporated into the sequence of a capsid protein (fiber, hexon, or pXI) constituting RAVNs likewise yielded encouraging results. Model experiments in mice demonstrated that administration of RAVNs harboring epitope sequences of influenza virus hemagglutinin within coat protein polypeptides elicited a pronounced humoral response in the animals [54]. Laboratory studies of the possibility of using target-specific RAVNs that interact with antigen-presenting (dendritic) cells as a vaccine are currently under way [55, 56].

The ability of RAVNs to induce T cell immune responses served as a rationale for the development of a new therapeutic approach (tumor vaccine therapy) and RAVN-based agents for this therapeutic approach. This approach is mechanistically very similar to vaccination against intracellular pathogens. Tumor cells produce specific antigens, presenting their epitopes on the surface as peptide fragments bound by MHC (major histocompatibility complex) class I molecules. Appropriate subsets of cytotoxic lymphocytes may detect and kill such cells. Administration of RAVNs expressing tumor antigens results (as in the case of vaccines against infectious agents) in proliferation of cytotoxic lymphocytes. Over fifty clinical trials (phase I and phase II) of such tumor vaccines are currently under way worldwide (http://www.wiley.co.uk/genmed/clinical/), exploring their therapeutic potential in patients

TABLE 5.2

Statistical Data Derived from Clinical Trials in the Field of Gene Therapy

Phase of Clinical Trials	Disease					
	Cancer	Cardiovascular	Infectious	Monogenic	Neurological	Other
I	169[a]/580[b]	25/59	7/63	16/85	1/7	4/31
I/II	44/177	6/30	0/27	2/32	0/5	0/4
II	43/163	15/36	2/16	0/3	0/5	1/12
II/III	6/7	4/6	0/0	0/0	0/0	0/0
III	9/33	2/6	0/6	0/0	0/0	0/2

Source: Journal of Gene Medicine. http://www.wiley.co.uk/genmed/clinical/. Used with permission.

[a] Number of clinical trials involving RAVNs.

[b] Total number of clinical trials.

with bladder cancer, melanoma, Hodgkin's lymphoma, nasopharyngeal carcinoma, chronic lymphoblastic lymphoma, prostate cancer, and other malignant diseases (Table 5.2).

RAVNs may well serve as a base for the future development of next-generation vaccines, the efficiency and safety of which will be warranted by an appropriate combination of genetic and physicochemical modification strategies. For example, such a vaccine may appear as a DNA molecule harboring an expression cassette with a gene of the infectious agent, which is packed into a synthetic (rather than viral) coat that

- Carries surface antibodies against the pathogen
- Prevents rapid elimination of RAVNs from the organism
- Facilitates specific interaction of RAVNs with and their entry into antigen-presenting cells

When developed, vaccines with such proteins will favor the formation of strong immune responses against the target pathogens.

5.8.3   Constructing RAVNs for Noninvasive Diagnostics

The potential of noninvasive diagnostics is quite considerable. It may serve to ensure early detection of a variety of diseases and facilitate studies of their course. Early detection of tumors and cardiovascular diseases, achieved by implementing methods of noninvasive diagnostics, is particularly important. Recent discovery of molecules and ligands, ensuring highly specific results of diagnostic tests, is currently paving the way to real-time studies of diverse events that take place at the molecular level in the course of disease development. Some of the methodologies in question have already been implemented in laboratory medicine (computer tomography, magnetic resonance, etc.).

RAVNs may become a promising platform for developing materials and methods to be used in noninvasive diagnostics. Diagnostic RAVNs are developed as a means of detecting particular types of cells in the organism. For example, target-specific RAVNs with one of the coat proteins modified by a fluorescent agent may be used for this purpose. Introducing such a modification produces uniformly labeled RAVNs. The first report of this strategy described the construction of human adenovirus sero-type 5–based RAVNs with the pIX gene modified to encode a green fluorescent protein (GFP)–pIX chimera [57]. GFP is used as a vital marker, which makes it possible to study multiple processes occurring within cells and organisms (otherwise not readily visualized, if at all). In vivo visualization of the results obtained involves the use of cooled charge-coupled device (CCD) imaging.

Target-specific RAVNs, the DNA molecules of which incorporate expression cassettes with genes encoding the probe (e.g., fluorescent protein), may also have diagnostic utility. It should be noted that the gene of the fluorescent protein may be controlled by either a constitutively active promoter (in which case all cells penetrated by RAVNs are detected) or its inducible counterpart (which allows detecting particular cell types as well defects within these cells). The method was validated in laboratory animals. The results of these experiments demonstrated its feasibility as a means of tumor visualization and tumor growth monitoring [58].

Yet another avenue of RAVN development for noninvasive diagnostics involves, in addition to specific detection of cells, delivery of therapeutic agents to these targets. This approach may be widely used in oncology as a means of ensuring selectivity of treatment (delivery of chemotherapy agents to transformed cells) combined with an instrument for monitoring its efficiency (noninvasive diagnostics). Tumor chemotherapy agents currently in use affect, in addition to malignant cells, healthy tissues and organs. RAVN preparations making it possible to overcome this problem are currently undergoing clinical trials.

For example, a clinical trial in patients with prostate cancer (stage T1c) is being held to explore the therapeutic efficacy of RAVNs, the DNA of which incorporates genes encoding two enzymes, i.e., herpes simplex virus thymidine kinase (HSV1-TK) and human sodium/iodide symporter (hNIS). The gene encoding hNIS serves as a visualization tool to assess the results of the therapy. Following administration of the radioactive substrate (sodium pertechnetate labeled with ^{99m}Tc), the symporter, which is expressed in cells penetrated by RAVNs, will transport ^{99m}Tc to the cell interior. The location of sodium pertechnetate in the organism is determined by single-photon emission computed tomography (SPECT). Thymidine kinase is an enzyme that phosphorylates thymidine, converting it into the corresponding monophosphate, which is further phosphorylated by other kinases to yield thymidine triphosphate, which is used in DNA synthesis.

RAVNs expressing HSV1-TK and hNIS are administered directly into the tumor, where they find the way to the cells, in which the harbored genes are expressed. Two days later, the patients receive orally the chemotherapy agent, ganciclovir, which is a selective substrate for HSV1-TK. (Cellular thymidine kinases are incapable of effecting its phosphorylation.) Following ganciclovir phosphorylation by HSV1-TK, its monophosphate undergoes additional phosphorylation by other intracellular enzymes, and the product this forms is incorporated into DNA, causing chain termination at the stage of cell division. Chain termination during DNA synthesis results in cell death. The key point in this approach is that only dividing cells will be killed.

Tumor cells are renowned for their increased proliferative potential. In this way, selective destruction of tumor cells is achieved. In order to visualize the results of the therapy (in which RAVNs encoding HSV1-TK and hNIS are therapeutic agents), sodium pertechnetate is administered intravenously to the patients immediately prior to the analysis. This clinical trial, involving RAVNs that harbor within their DNA both the therapeutic gene (HSV1-TK) and the gene for SPECT visualization (monitoring) of the outcome of the therapy (hNIS), demonstrated a decrease in the growth of malignant cells in patients with prostate cancer (stage T1c). It was also shown that the employed method of noninvasive diagnostics is both efficient and safe [59].

5.9 SUMMARY

One can enumerate the major properties typical of recombinant adenoviral nanoconstructions (RAVNs) regardless of the method used for their formation:

- RAVNs lack self-contained capacity for completing the life cycle (reproduction), typical of the initial natural adenoviruses, i.e., the administration of RAVNs into the body does not lead to the development of an infection process with typical symptoms.
- The genetic material of RAVNs (in whole or in part) is not integrated into the genome of the cells.
- RAVNs are entirely eliminated from the organism within a period of four–five weeks.

These characteristics of RAVNs demonstrate the high safety of such structures for humans and, consequently, the prospects of their application in medicine. This is verified by statistical data on the clinical trials of in the area of gene therapy. As Table 5.2 demonstrates, 25% of all such studies involve RAVNs.

To date, many different RAVNs have been created. A development of any new RAVN type is always strictly determined by the purpose of its further application. The combination of methods of physicochemical and genetic modification makes it possible to create increasingly efficient RAVNs for various medical needs. The creation of multifunctional RAVNs with a complex of properties combining the ability to both diagnose and treat diseases and monitor the course of the disease and its treatment efficiency may become possible in the future.

REFERENCES

1. Laemmli, U. K., J. R. Paulson, and V. Hitchins. 1974. Maturation of the head of bacteriophage T4. V: A possible DNA packing mechanism: In vitro cleavage of the head proteins and the structure of the core of the polyhead. *J. Supramol. Struct.* 2:276–301.
2. Yevdokimov, Yu. M., V. I. Salyanov, S. V. Semenov, and S.Skuridin, G. 2011. *DNA Liquid-Crystalline Dispersions and Nanostructures*, ed. Yu. M. Yevdokimov. Boca Raton, FL: CRC Press / Taylor & Francis Group.
3. Bouligand, Y., and V. Norris. 2001. Chromosome separation and segregation in dinoflagellates and bacteria may depend on liquid crystalline states. *Biochimie* 83:187–92.

4. Leforestier, A., and F. Livolant. 2009. Structure of toroidal DNA collapsed inside the phage capsid. *Proc. Natl. Acad. Sci. USA* 106:9157–62.

5. Leforestier, A., and F. Livolant. 2010. The bacteriophage genome undergoes a succession of intracapsid phase transitions upon DNA ejection. *J. Mol. Biol.* 396:384–95.

6. Singh, R., and K. Kostarelos. 2009. Designer adenoviruses for nanomedicine and nanodiagnostics. *Trends Biotechnol.* 27:220–29.

7. Harrop, R., J. John, and M. W. Carroll. 2006. Recombinant viral vectors: Cancer vaccines. *Adv. Drug Deliv. Rev.* 58:931–47.

8. Peek, L. J., C. R. Middaugh, and C. Berkland. 2008. Nanotechnology in vaccine delivery. *Adv. Drug Deliv. Rev.* 60:915–28.

9. Rodrigues, T., M. J. Carrondo, P. M. Alves, and P. E. Cruz. 2007. Purification of retroviral vectors for clinical application: Biological implications and technological challenges. *J. Biotechnol.* 127:520–41.

10. Stewart, P. L., S. D. Fuller, and R. M. Burnett. 1993. Difference imaging of adenovirus: Bridging the resolution gap between X-ray crystallography and electron microscopy. *EMBO J.* 12:2589–99.

11. Russell, W. C. 2009. Adenoviruses: Update on structure and function. *J. Gen. Virol.* 90:1–20.

12. Leppard, K. N. 2008. Adenoviruses: Molecular biology. In *Encyclopedia of Virology*, ed. B. W. J. Mahy and M. H. V. van Regenmortel, 17–23. Vol. 1, 3rd ed. Oxford, UK: Elsevier.

13. Freytag, S. O., H. Stricker, B. Movsas, and J. H. Kim. 2007. Prostate cancer gene therapy clinical trials (review). *Mol. Ther.* 15:1042–52.

14. Bischoff, J. R., D. H. Kirn, A. Williams, et al. 1996. An adenovirus mutant that replicates selectively in p53-deficient human tumor cells. *Science* 274:373–76.

15. Bett, A. J., W. Haddara, L. Prevec, and F. L. Graham. 1994. An efficient and flexible system for construction of adenovirus vectors with insertions or deletions in early regions 1 and 3. *Proc. Natl. Acad. Sci. USA* 91:8802–6.

16. Kochanek, S., G. Schiedner, and C. Volpers. 2001. High-capacity "gutless" adenoviral vectors. *Curr. Opin. Mol. Ther.* 3:454–63.

17. Fontana, L., M. Nuzzo, L. Urbanelli, and P. Monaci. 2003. General strategy for broadening adenovirus tropism. *J. Virol.* 77:11094–104.

18. Bouri, K., W. G. Feero, M. M. Myerburg, et al.1999. Polylysine modification of adenoviral fiber protein enhances muscle cell transduction. *Hum. Gene Ther.* 10:1633–40.

19. Chillon, M., A. Bosch, J. Zabner, et al. 1999. Group D adenoviruses infect primary central nervous system cells more efficiently than those from group C. *J. Virol.* 73:2537–40.

20. Li, Y., R. C. Pong, J. M. Bergelson, et al. 1999. Loss of adenoviral receptor expression in human bladder cancer cells: A potential impact on the efficacy of gene therapy. *Cancer Res.* 59:325–33.

21. Miller, C. R., D. J. Buchsbaum, P. N. Reynolds, et al. 1998. Differential susceptibility of primary and established human glioma cells to adenovirus infection: Targeting via the epidermal growth factor receptor achieves fiber receptor-independent gene transfer. *Cancer Res.* 58:5738–48.

22. Pickles, R. J., D. McCarty, H. Matsui, P. J. Hart, S. H. Randell, and R. C. Boucher. 1998. Limited entry of adenovirus vectors into well-differentiated airway epithelium is responsible for inefficient gene transfer. *J. Virol.* 72:6014–23.

23. Segerman, A., Y. F. Mei, and G. Wadell. 2000. Adenovirus types 11p and 35p show high binding efficiencies for committed hematopoietic cell lines and are infective to these cell lines. *J. Virol.* 74:1457–60.

24. Wickham, T. J., D. M. Segal, P. W. Roelvink, et al. 1996. Targeted adenovirus gene transfer to endothelial and smooth muscle cells by using bispecific antibodies. *J. Virol.* 70:6831–38.

25. Zabner, J., P. Freimuth, A. Puga, A. Fabrega, and M. J. Welsh. 1997. Lack of high affinity fiber receptor activity explains the resistance of ciliated airway epithelia to adenovirus infection. *J. Clin. Invest.* 100:1144–49.

26. Zhong, L., A. Granelli-Piperno, Y. Choi, and R. M. Steinman. 1999. Recombinant adenovirus is an efficient and nonperturbing genetic vector for human dendritic cells. *Eur. J. Immunol.* 29:964–67.

27. Barnett, B. G., C. J. Crews, and J. T. Douglas. 2002. Targeted adenovirus vectors. *Bioch. Bioph. Acta* 1575:1–14.

28. Logunov, D. Y., O. V. Zubkova, A. S. Karyagina-Zhulina, et al. 2007. Identification of HI-like loop in CELO adenovirus fiber for incorporation of receptor binding motifs. *J. Virol.* 81:9641–52.

29. Noureddini, S. C., and D. T. Curiel. 2005 Genetic targeting strategies for adenovirus. *Mol. Pharm.* 2:341–47.

30. Alemany, R., K. Suzuki, and D. T. Curiel. 2000. Blood clearance rates of adenovirus type 5 in mice. *J. Gen. Virol.* 81:2605–9.

31. Tao, N., G. P. Gao, M. Parr, et al. 2001. Sequestration of adenoviral vector by Kupffer cells leads to a nonlinear dose response of transduction in liver. *Mol. Ther.* 3:28–35.

32. Lozier, J. N., M. E. Metzger, R. E. Donahue, and R. A. Morgan. 1999. Adenovirus-mediated expression of human coagulation factor IX in the rhesus macaque is associated with dose-limiting toxicity. *Blood* 94:3968–75.

33. Hofherr, S. E., E. V. Shashkova, E. A. Weaver, R. Khare, and M. A. Barry. 2008. Modification of adenoviral vectors with polyethylene glycol modulates in vivo tissue tropism and gene expression. *Mol. Ther.* 16:1276–82.

34. Rihova, B. 2007. Biocompatibility and immunocompatibility of water-soluble polymers based on HPMA. *Compos. Part. B-Eng.* 38:386–97.

35. Lanciotti, J., A. Song, J. Doukas, et al. 2003. Targeting adenoviral vectors using heterofunctional polyethylene glycol FGF2 conjugates. *Mol. Ther.* 8:99–107.

36. Ogawara, K., M. G. Rots, R. J. Kok, et al. 2004. A novel strategy to modify adenovirus tropism and enhance transgene delivery to activated vascular endothelial cells in vitro and in vivo. *Hum. Gene Ther.* 15:433–43.

37. Eto, Y., J. Q. Gao, F. Sekiguchi, et al. 2005. PEGylated adenovirus vectors containing RGD peptides on the tip of PEG show high transduction efficiency and antibody evasion ability. *J. Gene Med.* 7:604–12.

38. Saini, V., D. V. Martyshkin, S. B. Mirov, et al. 2008. An adenoviral platform for selective self-assembly and targeted delivery of nanoparticles. *Small* 4:262–69.

39. Dorokhov, Y. L., M. V. Skulachev, P. A. Ivanov, et al. 2002. Polypurine (A)-rich sequences promote cross-kingdom conservation of internal ribosome entry. *Proc. Natl. Acad. Sci. USA* 99:5301–6.

40. Huang, W., and S. J. Flint. 1998. The tripartite leader sequence of subgroup C adenovirus major late mRNAs can increase the efficiency of mRNA export. *J. Virol.* 72:225–35.

41. Sheay, W., S. Nelson, I. Martinez, T. H. Chu, S. Bhatia, and R. Dornburg. 1993. Downstream insertion of the adenovirus tripartite leader sequence enhances expression in universal eukaryotic vectors. *Biotechniques* 15:856–62.

42. Tutykhina, I. L., M. M. Shmarov, D. Iu. Logunov, et al. 2008. Construction of the vector based on the CELO avian adenovirus genome providing enhanced expression of secreted alkaline phosphatase gene in a non-permissive system in vitro and in vivo. *Molecular Genetics, Microbiology and Virusology* (Russian ed.) 4:26–30.

43. Dai, Y., E. M. Schwarz, D. Gu, W. W. Zhang, N. Sarvetnick, and I. M. Verma. 1995. Cellular and humoral immune responses to adenoviral vectors containing factor IX gene: Tolerization of factor IX and vector antigens allows for long-term expression. *Proc. Natl. Acad. Sci. USA* 92:1401–5.

44. Dedieu, J. F., E. Vigne, C. Torrent, et al. 1997. Long-term gene delivery into the livers of immunocompetent mice with E1/E4-defective adenoviruses. *J. Virol.* 71:4626–37.

45. Bachtarzi, H., M. Stevenson, and K. Fisher. 2008. Cancer gene therapy with targeted adenoviruses. *Expert Opin. Drug Deliv.* 5:1231–40.

46. Subr, V., L. Kostka, T. Selby-Milic, et al. 2009. Coating of adenovirus type 5 with polymers containing quaternary amines prevents binding to blood components. *J. Control. Release* 135:152–58.

47. Tutykhina, I. L., O. A. Bezborodova, L. V. Verkhovskaia, et al. 2009. Recombinant pseudoadenovirus nanostructure with human lactoferrin gene: Production and study of lactoferrin expression and properties in vivo usage of the construction. *Molecular Genetics, Microbiology and Virusology* (Russian ed.) 1:27–31.

48. Garber, K. 2006. China approves world's first oncolytic virus therapy for cancer treatment. *J. Natl. Cancer Inst.* 98:298–300.

49. Guo, J., and H. Xin. 2006. Chinese gene therapy: Splicing out the West? *Science* 314:1232–35.

50. Zavalishin, I. A., N. P. Bochkov, Z. A. Suslina, et al. 2008. Genetherapy of amyotropic lateral sclerosis. *Bull. Exp. Biol. Med.* 145:483–86.

51. Avdeeva, S. V., D. A. Voronov, N. V. Khaidarova, et al. 2004. The CELO-ANG recombinant avian adenovirus with human angiogenine gene inducing neovascularization in the anterior tibial muscle of rat. *Molecular Genetics, Microbiology and Virusology* (Russian ed.) 4:38–40.

52. Bochkov, N. P., B. A. Konstantinov, A. V. Gavrilenko, et al. 2006. The technologies of genetic engineering in treatment of chronic lower limb ischemia. *Herald of the Russian Academy of Medical Sciences* (Russian ed.) 10 (9): 6–10.

53. Van Kampen, K. R., Z. Shi, P. Gao, et al. 2005. Safety and immunogenicity of adenovirus-vectored nasal and epicutaneous influenza vaccines in humans. *Vaccine* 23:1029–36.

54. Krause, A., J. H. Joh, N. R. Hackett, et al. 2006. Epitopes expressed in different adenovirus capsid proteins induce different levels of epitope-specific immunity. *J. Virol.* 80:5523–30.

55. Tian, Y., H. H. Zhang, L. Wei, et al. 2007. The functional evaluation of dendritic cell vaccines based on different hepatitis C virus nonstructural genes. *Viral Immunol.* 20:553–61.

56. Cheng, C., J. G. Gall, W. P. Kong, et al. 2007. Mechanism of ad5 vaccine immunity and toxicity: Fiber shaft targeting of dendritic cells. *PLoS Pathog.* 3:239–45.

57. Meulenbroek, R. A., K. L. Sargent, J. Lunde, B. J. Jasmin, and R. J. Parks. 2004. Use of adenovirus protein IX (pIX) to display large polypeptides on the virion-generation of fluorescent virus through the incorporation of pIX-GFP. *Mol. Ther.* 9:617–24.

58. Ono, H. A., L. P. Le, J. G. Davydova, T. Gavrikova, and M. Yamamoto. 2005. Noninvasive visualization of adenovirus replication with a fluorescent reporter in the E3 region. *Cancer Res.* 65:10154–58.

59. Barton, K. N., H. Stricker, S. L. Brown, et al. 2008. Phase 1 study of noninvasive imaging of adenovirus-mediated gene expression in the human prostate. *Mol. Ther.* 16:1761–69.

6 Application of "Liquid" and "Rigid" DNA Nanoconstructions Immobilized in Polymeric Hydrogel as Sensing Units

6.1 ISSUES OF STABILIZATION OF PARTICLES OF LIQUID-CRYSTALLINE DISPERSIONS

Regardless of the possible advantages of various sensing units based on DNA cholesteric liquid-crystalline dispersions (CLCD) or DNA nanoconstructions (NaC), the principal difficulty in the application of the "test-tube" form of such units is the insufficient temporal stability of their optical properties. This instability is caused by the fact that at a long storage (longer than a week), as the result of the sedimentation and/or coalescence of CLCD particles or DNA NaC, their concentration in the solution is diminished. Such change in concentration results in an uncontrollable alteration of the optical signal generated by these nanoobjects.

The problem of stabilization of the optical properties of liquid crystals of various compounds is discussed in the literature. This issue obtains especially significant meaning in the case of cholesteric liquid crystals and dispersions of any compounds that have the most mobile spatial structure, as discussed in chapters 3 and 4.

Two approaches to the stabilization of the abnormal optical properties of LCD particles have been described. In the case of LCD particles of low-molecular-mass compounds ("LC drops") that usually exist in organic solvents and are used in technology, the goal of the stabilization is not only the maintenance of the abnormal optical activity, but also the protection of the molecules forming the LC from the damaging effects of the environment (humidity, etc.). It is obvious that the sedimentation and/or aggregation of dispersion particles of any compounds can be prevented by somehow fixing the spatial location of these particles, in particular, by creating a specific network or wall around the particles. This goal is achieved by creating synthetic films of polymers with low-molecular-mass LCD particles embedded into them as isolated phase drops [1]. The method of phase separation is frequently applied to receive films with LC drops immobilized in their content. The method includes the exclusion of the phase enriched with low-molecular-mass cholesteric in a form of isolated drops from the phase capable of forming the film by a change in the temperature or the pH of the solution, evaporating the solvent, or adding a flocculent to the solution. Natural (gelatin, cellulose and its derivatives) and synthetic (polyvinyl alcohol, polyvinyl acetate, polyvinylchloride, polysiloxanes, epoxy-resins) polymers were used as a film-forming material.

The drops of the cholesteric LC of low-molecular-mass compounds in polyvinyl alcohol films were first realized in 1968 [2]. In the following years, patents specifying and developing the methods of stabilization of the drops in the films [3] and proposing other film-forming polymers [4–6] appeared.

For the stabilization of microscopic drops of low-molecular cholesterics in polymeric films, the polymer must fit the following requirements.

1. The polymer must not form complexes with the compounds forming the cholesteric.
2. The polymer must be transparent in the required region of the spectrum.
3. The polymer must have a set of certain physicomechanical properties, namely, elasticity at a broad range of temperatures, surface tension that determines the necessary orientation of the cholesteric molecules inside the drops, and a refractive coefficient close to the average refractive coefficient of the cholesterics ($n = 1.48–1.58$).

Theoretically, the structures existing in the cholesteric drops, suspended and isotropic fluids, whose properties are determined by the limiting conditions on the cholesteric-isotropic border, must be realized in the formed films [7, 8].

The formation of films containing LC drops of synthetic polypeptides (polyamino acids) in polymeric films received by the polymerization of butylacrylate is described in the literature [9, 10]. Such films have high stiffness determined by the high density of the polymeric network. However, the polymerization of butylacrylate takes place under such conditions that can distort the spatial structure of the synthetic polypeptides immobilized within the film.

Indeed, in practice the optical properties of the LC drops in polymeric films can be different from the properties of the initial LC. Changes in the spectrum of the reflected light can be noticed: The reflection peaks are broadened; the intensity of the peaks is reduced; and the dependence of the wavelength of the reflected light on the temperature changes. Consequently, a polymeric film (matrix) exerts an effect on the properties of the LC drops [11]. Such an effect is caused by the large mechanical tensions that evolve in the films during the process of formation (in particular, as the result of the evaporation of the solvent). It is obvious that such tensions distort the tangential orientation of the molecules forming the cholesteric phase [12].

To restore the ordered initial structure of the drops, a thermal shock (thermal training) is used, i.e., the films are kept at elevated temperature. This process significantly improves the optical properties of the films containing the LC polymeric drops [13]. Therefore, approaches to the inclusion of LC drops of low-molecular-mass compounds into rigid polymeric films that can conserve their optical properties have been developed. The high density of the polymeric network makes the films almost impenetrable for the solvent molecules. At the same time, the advantages of the films are the ability to take any shape and their unlimited size; the films also can combine rigidity and elasticity.

However, in the case of DNA cholesteric LCD (CLCD) particles and DNA NaCs existing in water–salt solutions at a certain osmotic pressure, completely different goals of stabilization must be achieved. These goals are not only the preservation of the abnormal optical activity, but also the creation of the conditions at which the low-molecular-mass compounds can diffuse into the film and interact with the DNA molecules. Consequently, at the practical application of dispersion particles for analytical purposes, the key role belongs to the properties of the polymeric materials capable of including the DNA CLCD particles (DNA NaC). The material must comply with a number of requirements:

1. Must possess high temporal stability at storage
2. Must possess chemical and biological inertness
3. Must not affect the physicochemical properties of the LCDs of biopolymers, namely, DNA
4. Must not possess a significant optical absorption in the region of wavelength from 220 to 750 nm
5. Must not possess an optical isotropy in the wavelength region from 220 to 750 nm
6. Must provide the conditions for the diffusion of chemically relevant or biologically active molecules without changing their parameters
7. Must allow registering the properties of the biopolymer LCDs with optical methods, namely, measurement of CD spectra via spectroscopy of circular dichroism (CD)

The enumerated requirements can be followed if the polymeric material in which the isolated biopolymer molecules and their LC are immobilized is not an impenetrable film, but a network spatial structure known as a polymeric gel or polymeric matrix. Taking into account the necessity to study the biopolymers in water–salt solutions and considering the enumerated requirements, the technology of the production of polymeric hydrogels is of great importance.

One of the most popular methods of obtaining polymeric hydrogels containing isolated protein molecules based on the polymerization of a hydrophilic monomer—acrylamide molecules added to the solutions of these proteins—was first described in 1963 in the work of Bernfeld and Wan [14].

6.2 IMMOBILIZATION OF DNA LCD PARTICLES IN A HYDROGEL

The first attempt to create a polymeric hydrogel containing DNA CLCD particles (as well as of poly(I)×poly(C)) fixed in its structure was taken in the work of Fukui et al. [15] and based on the polymerization of acrylamide introduced into the solutions of poly(ethylene)glycol (PEG) or a PEG-copolymer with poly(propylene glycol), i.e., pluronic, with CLCD particles preformed in these solution. This approach led to the formation of a polymeric hydrogel with CLCD particles immobilized in its structure, and their CD spectra were only slightly different from the CD spectra of the initial CLCD of DNA or poly(I)×poly(C). However, the low transparency of the hydrogel determined by the partial incompatibility of PEG (or pluronic) with poly(acrylamide), as well as insufficient chemical and physical stability of CLCD in the structure of this kind of polymeric hydrogel, did not provide the opportunity to use the hydrogels with CLCD particles as stabilized sensing elements.

The approach that proved to be more convenient for the creation of polymeric hydrogels with controllable structure is based on the application of the water–salt solutions of so-called PEG-macromonomers, i.e., PEG molecules containing end groups capable of radical polymerization. In particular, PEG-macromonomers, for instance, PEG-dimethylacrylates, have a common formula:

$$CH_2=C(CH_3)-C(=O)-O-(-CH_2-CH_2-O-)_n-C(=O)-C(CH_3)=CH_2$$

Such macromonomers have a molecular mass of approximately 1,000. The radical polymerization of PEG-macromonomers initiated by the UV-irradiation (wavelength 360 nm, irradiation time 3 min) leads to the formation of a polymeric hydrogel with enzyme (invertase) molecules immobilized in its structure, and the hydrogel was penetrable for the glucose molecules [16].

The most successful approach was based on the consideration of properties of both PEG and acrylamide molecules and proposed in the work of Kazanskii et al. [17]. As shown in chapters 3 and 4, the DNA CLCD particles formed in a water–salt solution of PEG exist at least for some period of time after the formation as independent particles. At the same time, the concentration of neighboring PEG molecules excluded from the CLCD particles and surrounding the particles is quite high. The idea of Kazanskii et al. [17] was very simple: to cross-link the neighboring PEG molecules without changing their spatial positions. In this case, a polymeric network (polymeric matrix) with preformed CLCD particles fixed in "cells" of this matrix would be formed. If the created polymeric matrix does not influence the spatial structure of the DNA CLCD particles, the abnormal optical activity of the particles can be retained (Figure 6.1). (In Figure 6.1, panel A depicts PEG used for the formation of a DNA CLCD and the CD spectrum of DNA CLCD; panel B depicts the PEG-macromonomer used for the formation of DNA CLCD and the CD spectrum of solution containing DNA CLCD in PEG-macromonomer; panel C is the cross-linking of PEG-macromonomers that does not distort the structure of DNA CLCD particles and must only lead to a minor change in the amplitude of the abnormal band in the CD spectrum. The X marks the rotation axis of the cholesteric formed.)

The obvious advantage of this approach to the stabilization of DNA CLCD particles is the opportunity to perform the reaction of synthesis of the polymeric matrix in a water–salt solution. Two problems had to be solved in order to implement this approach: (a) synthesizing the PEG-macromonomer (or PEG-macromonomers) such that it did not affect the condition required for the phase exclusion of DNA molecules and the optical activity of the formed CLCD particles (Figure 6.1(B)) and (b) finding a convenient method of polymerizing the macromonomer while retaining the abnormal optical activity of the particles.

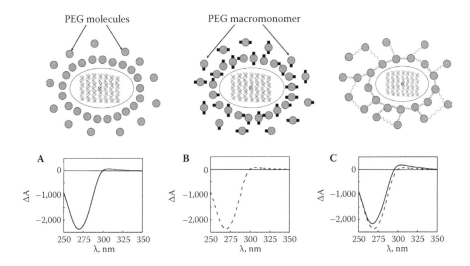

FIGURE 6.1 Hypothetical scheme illustrating the properties of DNA CLCD particles in water–polymer solutions and in a polymer hydrogel.

A PEG molecule with two end OH-groups allows multiple functionalization reactions. The presence of these groups makes it possible to receive both mono- and bis-macromonomers of PEG with deliberately selected reactive groups with a high yield, which is necessary for the three-dimensional (3-D) polymerization. In order to synthesize mono- and bis-macromonomers, namely, derivatives of PEG-bis-metacrylate, PEG preparations available at a broad range of molecular masses (from 1,000 to 40,000 g/mol) were used. The application of PEG-bis-metacrylate derivatives is based on the fact that they have a high reactivity at moderate temperatures. The functionality of PEG monomers, their molecular mass, and the conditions of the polymerization reaction can be selected so that a highly elastic material with a structure (a) that provides both soft fixation of DNA CLCD particles without affecting their spatial structure for a long period of time and (b) that is convenient for the diffusion of low-molecular-mass chemical and biologically active compounds without changing their initial structure and properties as the result of the polymerization. At the same time, the functionality, i.e., the average number of metacrylate groups per a PEG monomer must be within the interval between 1 and 2, which, on the one hand, provides a low concentration of the initial PEG and the quick attainment of the point of gel formation and, on the other hand, provides the specified density of the polymeric network and its penetrability for outside compounds.

PEG-macromonomers with molecular mass 4,000 and 6,000 g/mol were synthesized. The initial PEG preparation and the received mono- and bis-macromonomers were nontoxic, chemically neutral to DNA, and had a high solubility in water–salt solutions with physiological ionic strength necessary for the phase exclusion of double-stranded DNA molecules. At the same time, the water–salt solutions of PEG-macromonomers were optically isotropic and had sufficient transparency for the registration of their absorption and CD spectra.

DNA CLCD particles with the CD spectra almost identical to the CD spectra of CLCD formed from these molecules in the solutions of the initial PEG preparation were formed in the solutions of these macromonomers. At the same time, the "border" conditions necessary of the formation of DNA CLCD particles in water–salt solutions of PEG-macromonomers were almost practically identical to that in the solutions of the initial PEG preparations.

The polymerization of PEG-macromonomer in solutions containing preformed DNA CLCD particles was induced by persulfate initiators ($Na_2S_2O_8$-$Na_2S_2O_3$) at 30°C. The three-dimensional radical polymerization of PEG-macromonomer in a water–salt solution in the presence of DNA CLCD particles with controllable properties led to the formation of hydrogels with individual CLCD particles fixed in their structure. After the polymerization, the CD spectra of DNA CLCD particles formed in the initial PEG solution were compared to the spectra of the CLCD particles in the PEG-macromonomer solution and in the polymeric hydrogel (Figure 6.2). (In Figure 6.2, curve 1 is the CD spectrum of a DNA CLCD formed in a water–salt solution of PEG; C_{PEG} = 170 mg/mL; PEG molecular mass = 6,000 g/mol; 0.3 mol/L NaCl + 10^{-2} phosphate buffer; curve 2 is the CD spectrum of a DNA CLCD formed in a water–salt solution of PEG-macromonomer; curve 3 is the CD spectrum of DNA CLCD in a hydrogel obtained by the polymerization of PEG-macromonomer.)

It is worth recalling that a minor difference between the abnormal optical activity of DNA CLCD particles in a polymeric hydrogel and the abnormal optical activity of the same particles in the PEG-macromonomer solution may indicate that, at the transition from the solution to the hydrogel, only a small decrease in the osmotic pressure takes place, influencing the properties of the DNA CLCD particles, as discussed in chapters 3 and 4. Therefore, a polymeric hydrogel (polymeric matrix) based on PEG- metacrylate-macromonomer basically does not distort the spatial structure of DNA CLCD particles.

A hydrogel can have almost any shape convenient for the experiments. A hydrogel containing DNA CLCD retains its abnormal optical activity at least for six months from the moment of its preparation if relatively mild conditions (humidity,

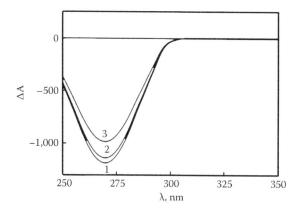

FIGURE 6.2 Experimentally measured CD spectra of DNA CLCD in polymer solutions and in a hydrogel.

temperature) are maintained. This means that a hydrogel is a chemically stable and biologically inert material. The peculiarity of a hydrogel is that, due to the change of the molecular mass and the combination of the concentrations of mono- and bis-macromonomers, the density of the forming polymeric matrix and the rate of the diffusion of compounds with different molecular mass into the hydrogel can be regulated.

It is necessary to emphasize that from the moment of the formation of a covalently cross-linked hydrogel, the DNA molecules forming each of the isolated CLCD particles cannot leave its volume and behave as a united unit, changing only its own volume and, consequently, the type of the spatial packing in response to the changes in the external conditions, namely, the osmotic pressure. Indeed, if a piece of a hydrogel with a fixed size is placed into a water–salt solution of the analyzed compound, two independent processes take place simultaneously.

First, the hydrogel swelling that takes place leads to an increase in its volume and, consequently, to a reduction in the local PEG concentration and a drop (up to a total disappearance) in the abnormal optical activity of the DNA CLCD particles, as seen in Figure 6.3. (In Figure 6.3, panel A shows the initial hydrogel and the CD spectrum of DNA CLCD in the hydrogel below; panel B shows the hydrogel after swelling and the CD spectrum of DNA CLCD below.) It must be noted that after further osmotic compression of the polymeric hydrogel, the abnormal optical properties of the CLCD particles are completely restored. This verifies the fact that the formation

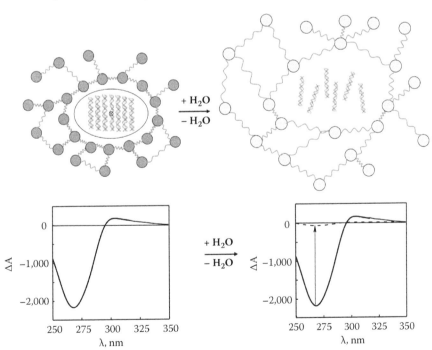

FIGURE 6.3 Hypothetical scheme illustrating the properties of DNA CLCD particles in a polymeric hydrogel after its swelling.

of a hydrogel does not affect the type of the spatial packing of DNA molecules in the CLCD particles.

Second, "outside" molecules (for instance, biologically active compounds, BACs) diffuse into the hydrogel and interact with DNA. If BAC has an absorption band and the BAC molecules are fixed at a certain angle to the axes of DNA molecules, this interaction is followed by the appearance of an abnormal optical activity in the CD spectrum in the region of absorption of BAC chromophores.

The combination of these two different processes can be mathematically described, although this makes the plotting of the calibration curves more difficult.

However, if a piece of the hydrogel is placed into a water–salt PEG-containing solution with a concentration that corresponds to the initial concentration of PEG macromonomer (supporting solution), the swelling of the hydrogel hardly occurs, though the diffusion of BAC molecules into the hydrogel is retained.

By observing the results obtained using individual representatives of various groups of colored biologically active compounds (dyes, antibiotics, antitumor drugs), it was shown that these compounds can diffuse into a hydrogel and react with DNA molecules in the content of CLCD particles, thus causing changes in the CD spectra typical of these compounds.

Figure 6.4 shows the CD spectra of the DNA CLCD in a hydrogel placed in a PEG-containing "supporting" solution with added daunomycin (DAU). (In Figure 6.4, curve 1 shows results for 30 min; curve 2, 60 min; curve 3, 90 min; curve 4, 120 min; curve 5, 150 min. Conditions: C_{DNA} = 30 μg/mL; C_{PEG} = 170 mg/mL; C_{DAU} = 1.7 × 10^{-5} mol/L; PEG molecular mass is 40,000 g/mol; 0.3 mol/L NaCl + 10^{-2} mol/L phosphate buffer; hydrogel sample size is 6 × 3 × 20 mm; ΔA = $(A_L - A_R) \times 10^{-6}$ optical units.)

The intense negative band with a maximum at $\lambda \approx 500$ nm, whose shape is similar to that of the DAU absorption band, indicates that DAU molecules diffuse into the hydrogel and form a complex with DNA molecules forming the CLCD. The higher

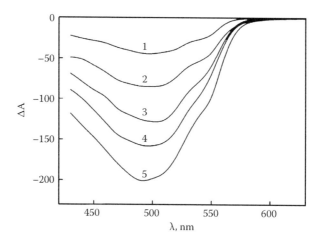

FIGURE 6.4 CD spectra (visible region) of the DNA CLCD in a hydrogel placed in PEG-containing supporting solution with daunomycin recorded within different time intervals.

FIGURE 6.5 (**See color insert**) Piece of a polymeric hydrogel containing DNA CLCD particles pretreated with SYBR Green fluorescent dye.

the concentration of DAU molecules bound to DNA, the greater the amplitude of the band at $\lambda \approx 500$ nm.

An additional example is shown in Figure 6.5. (The inset in Figure 6.5 is an enhanced image of a hydrogel fragment. Images obtained from a Leica DMI 4000 B fluorescent microscope.) A hydrogel was placed into a PEG-containing supporting solution containing the fluorescent SYBR Green dye. Then, with the help of a fluorescent microscope, the images of the DNA CLCD particles in this solution were taken. Figure 6.5 shows the image of the piece of the hydrogel sample where the shining points correspond to the CLCD particles with SYBR Green dye molecules fixed between the DNA base pairs as the result of intercalation. It can be seen that the DNA CLCD particles exist as independent objects located at a significant distance from each other.

Moreover, the CD spectrum of the hydrogel after keeping it in the PEG-containing solution of SYBR Green dye indicates the existence of two abnormal bands, as seen in Figure 6.6. (In Figure 6.6, curve 1 is the initial polymeric hydrogel; curve 2 is the polymeric hydrogel incubated with SYBR Green dye. Conditions: $C_{PEG} = 170$ mg/ mL; PEG molecular mass = 4,000 g/mol; 0.3 mol/L NaCl + 10^{-2} mol/L phosphate buffer; sample size is 7×3×25 mm.) One of the bands is located in the region of DNA chromophore absorption ($\lambda \approx 270$ nm), and the other one lies in the region of SYBR Green dye chromophore absorption ($\lambda \approx 500$ nm). First, these results completely correspond to the concept of the intercalation of SYBR Green molecules between the DNA base pairs in CLCD particles fixed in the spatial structure of a hydrogel. Second, the results show that molecules of low-molecular-mass compounds can both diffuse into a hydrogel and interact with DNA molecules in CLCD particles.

Figures 6.5 and 6.6 demonstrate that a hydrogel based on PEG-macromonomer provides the conditions for the diffusion of outside molecules in their initial, intact

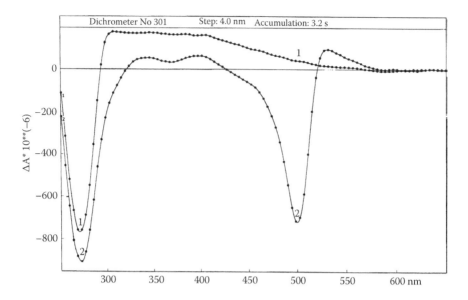

FIGURE 6.6 Experimental CD spectra of a polymeric hydrogel containing DNA CLCD particles, and the same hydrogel after its incubation in a supporting solution of PEG containing SYBR Green fluorescent dye.

state. The results, given in Figures 6.5 and 6.6, mean that a hydrogel can be used as a convenient system for investigating the mode of ordering DNA molecules in CLCD particles. Moreover, in the case where a supporting PEG solution containing the molecules of the analyzed substance (the analyte) is used, the hydrogel can be used as a sensing element. However, in this case, the application of an additional PEG-containing supporting solution creates certain difficulties for the researcher and, consequently, the exact calculation of the concentration of BACs in the analyzed solution becomes more difficult.

6.3 FORMATION AND IMMOBILIZATION OF DNA NANOCONSTRUCTION IN A HYDROGEL

An attempt was made to form a NaC based on the DNA molecules in CLCD particles fixed in the spatial structure of a hydrogel [18] using a technique described in chapters 3 and 4. The idea of the formation of a NaC under such conditions attracts both fundamental and practical interest. The matter is that a NaC can exist in the absence of a high osmotic pressure of a PEG-containing solution. Consequently, even after swelling of the hydrogel and a drop in the local PEG concentration, the abnormal optical activity typical of NaCs must be retained as the hydrogel is placed into a water–salt solution. The abnormal optical activity can only disappear if the nanobridges formed between the neighboring DNA molecules in the content of NaCs are destroyed under the effect of the analytes.

To form a NaC directly in the structure of a polymeric hydrogel, the hydrogel in which the DNA CLCD particles were formed was treated with solutions of DAU and copper, as described in chapters 3 and 4. In Figure 6.7, the CD spectrum of the initial

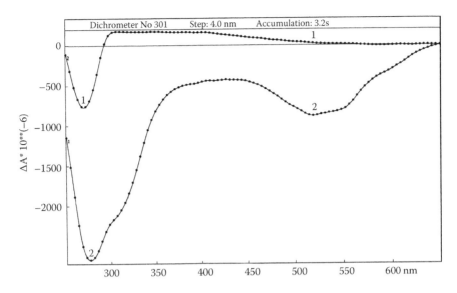

FIGURE 6.7 Experimental CD spectra of a polymeric hydrogel containing DNA CLCD particles and DNA NaCs formed from these particles.

DNA CLCD in a hydrogel is compared to the CD spectrum of NaCs formed by the described technique. (In Figure 6.7, curve 1 shows the DNA CLCD in the initial hydrogel; curve 2 shows the DNA NaCs formed from CLCD particles fixed in the structure of the polymeric hydrogel. Sample size is 7×3×25 mm.) The amplification of the abnormal optical activity in the absorption region of the DNA chromophores and the appearance of the second band in the visible region of the spectrum (the absorption of DAU chromophores) testifies to the formation of a NaC that is based on the DNA molecules initially fixed in the spatial structure of CLCD particles and stabilized in a polymeric hydrogel.

It should be noted that the temporal stability of a NaC in a hydrogel is quite high, with the amplitude of the abnormal band typical of the DNA NaCs typically being retained for at least six months. This result is of great practical importance because it opens the way for the practical application of a new type of NaCs.

The combination of physicomechanical properties of the hydrogel and the presence of the abnormal band in the CD spectrum (easily detectable with the help of a portable dichrometer) makes it possible to use the hydrogel based on NaCs as a biosensing unit for BACs interacting with both the DNA molecules and the elements of the nanobridges. As an example, a hydrogel containing NaCs was placed into a solution of homocystein (Hc). The homocystein molecules diffuse into the hydrogel and extract the copper ions from the nanobridges between the DNA molecules in the NaC structure; this process is followed by a decrease in the amplitude of the abnormal band in the CD spectrum.

Figure 6.8 shows the CD spectra in the visible region of the spectrum (the absorption region of DAU chromophores) measured at different periods of time after the addition of homocystein. (In Figure 6.8, curve 1 is 0 min; curve 2, 25 min; curve 3,

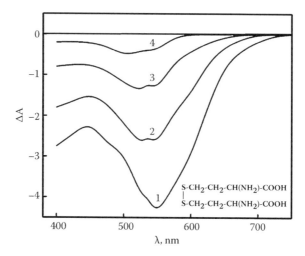

FIGURE 6.8   CD spectra of DNA NaCs in a polymeric hydrogel after its swelling in a solution containing homocystein registered after different periods of time.

45 min; curve 4, 115 min. Conditions: $C_{DNA} = 20$ μg/mL; $C_{PEG} = 120$ mg/mL; PEG molecular mass = 35,000 g/mol; 0.3 mol/L NaCl + 10^{-2} mol/L phosphate buffer; $C_{Hc} = 39.2$ μg/mL; thickness of the hydrogel sample = 3 mm; $\Delta A = (A_L - A_R) \times 10^{-3}$ optical units. The structural formula of homocystein is given in the right corner of the figure.) The decrease in the band amplitude in this case reflects the disintegration of the nanobridges in the DNA NaC structure.

The curves reflecting the destruction of DNA NaCs in a polymeric hydrogel under the effect of compounds of different molecular mass—namely, homocystein (135 g/mol) and heparin (4,000 g/mol) diffusing into the hydrogel at different rates—are compared in Figure 6.9. It should be noted that the efficiency of the destruction of DNA NaCs must be related both to the diffusion coefficients of the compounds diffusing into the polymeric hydrogel and the size of the cells in the polymeric network. The latter parameter can be regulated by selecting the molecular mass of PEG-macromonomers and/or the number of metacrylate groups in the macromonomers.

Comparison of all of these results indicates that the DNA CLCD particles or DNA NaCs can be conserved in the structure of polymeric hydrogels without alteration of their abnormal optical properties. At the same time, the method for the creation of polymeric hydrogels is relatively simple and can be carried out at room temperature using a water–salt solution. Apparently, this method can be applied to preserve other biological structure and objects.

6.4   SUMMARY

The physicochemical properties of a transparent, optically isotropic, polymeric hydrogel (polymeric matrix) based on derivatives of PEG metacrylate both (a) provides the stabilization immobilization of DNA CLCD particles and DNA NaC with a size about 500 nm and (b) opens the opportunity to include CLCD particles of both

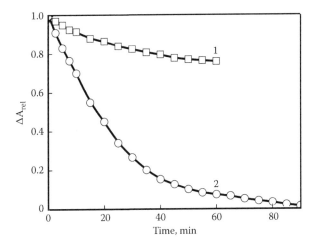

FIGURE 6.9 Curves of the destruction of DNA NaC in a polymeric hydrogel under the effect of homocystein (curve 1) and heparin (curve 2).

nucleic acids and their complexes with various BACs into a hydrogel. The immobilization of liquid and rigid DNA NaCs in the structure of polymeric hydrogels both preserves the abnormal optical activity of these objects and provides the conditions for retaining the reactivity of DNA molecules in such structurally sensitive objects as DNA NaCs.

The properties of the hydrogels provide an efficient diffusion of BACs of different molecular mass into the hydrogel. These properties minimize the probable parasitic optical effects at the measurement of the abnormal optical activity that shows itself at the interaction of BACs with DNA molecules, increase the accuracy of the measurements of the abnormal optical activity, and broaden the areas of possible application of the created polymer hydrogel.

The polymeric hydrogels containing DNA NaCs can be used as multifunctional sensing units in test systems designed for the screening of BACs and such compounds as antitumor drugs, antibiotics, toxins, etc., whose targets are both elements of the nanobridges and the genetic material of a cell. In this case, a biodetector is a synthetic polymeric matrix containing spatially separated DNA CLCD particles and DNA NaCs. The principle behind the functioning of such biodetectors is similar to that of film-type indicators. Moreover, the DNA NaCs with controllable physicochemical properties immobilized in the content of polymeric gel can be used in technology (in particular as molecular sieves, optical filters, and so on) and in various application fields, including biosensorics, optics, and electronics.

REFERENCES

1. Zharkova, G. M., and A. S. Sonin. 1994. *Liquid Crystalline Composites* (Russian ed.). Novosibirsk, Russia: Nauka.
2. Churchill, R. D., J. V. Carmel, and R. E. Miller. *Visual display device*. UK Patent 1,161,039, Publ. 68.

3. Zozzi, Z. A. 1970. Microencapsulation. *J. Pharm. Sci.* 59:1367–69.

4. Churchill, R. D., and J. V. Carmel. *Display device containing minute droplets of cholesterics in a substantially continuous polymeric matrix.* US Patent 3,600,060, Publ. 71.

5. Hesse, R., G. Edler, and H. Keller. *Liquid crystal-based laquers.* DE Patent 2,201,121, Publ. 72.

6. Churchill, R. D., J. V. Carmel, and R. E. Viller. *Gelatin-gum arabic capsules containing cholesteric liquid crystal material.* US Patent 3,697,297, Publ. 72.

7. Kurik, M. V., and O. D. Lavrentovich. 1982. Negative-positive monopole transitions in cholesteric liquid crystals. *JETP Lett.* (Russian ed.) 35:362–65.

8. Zharkova, G. M., and S. I. Traksheyev. 1989. The orientation of liquid crystals in a spherical volume. *Crystallography Reports* (Russian ed.) 34:695–701.

9. Tsutsui, T., and R. Tanaka. 1980. Immobilization of cholesteric structure in poly(γ-butyl L-glutamate)-butyl acrylate liquid crystalline system. *J. Polym. Sci. Polym. Lett. Ed.* 18:17–23.

10. Tsutsui, T., and R. Tanaka. 1981. Network polymers with cholesteric liquid crystalline order prepared from poly(γ-butyl L-glutamate)-butyl acrylate liquid crystalline system. *Polymer* 22:117–23.

11. Zharkova, G. M., V. P. Mamaev, E. P. Fokin, V. M. Khachaturyan, and A. I. Shchadrina. Optical properties of the polymer–liquid crystals–pyrimidine derivatives system. *Bulletin of the Siberian Branch of the USSR Academy of Sciences, a chemical series* (Russian ed.) 8:3–6.

12. Zharkova, G. M. 1988. The development of crystallographic thermography as applied to problems of heat transfer. PhD diss., Novosibirsk State Univ.

13. Sonin, A. S., and V. P. Shibaev. 1981. Structural ordering and properties of cholesteric pseudocapsulated films. *Russian Journal of Physical Chemistry A* (Russian ed.) 55:1263–68.

14. Bernfeld, P., and J. Wan. 1963. Antigens and enzymes made insoluble by entrapping them into lattice of synthetic polymers. *Science* 142:678–79.

15. Fukui, S., A. Tanaka, T. Iida, and E. Hasegawa. 1976. Application of photo-crosslinkable resin to immobilization of an enzyme. *FEBS Lett.* 66:179–82.

16. Yevdokimov, Yu. M., and S. G. Skuridin. 1988. Pseudocapsulated liquid crystals of nucleic acids. *Doklady of the USSR Academy of Sciences* (Russian ed.) 303:232–35.

17. Kazanskii, K. S., S. G. Skuridin, V. I. Kuznetsova, and Yu. M. Yevdokimov. 1996. Poly(ethylene oxide) hydrogels containing immobilized particles of an LC dispersion of deoxyribonucleic acid. *Polymer Science, Series A* (Russian ed.) 38:875–83.

18. Kazanskii, K. S., M. A. Lagutina, S. G. Skuridin, and Yu. M. Yevdokimov. 2007. Nanoconstructions based on double-stranded DNA in poly(ethylene oxide) hydrogels. *Doklady of the Russian Academy of Sciences* (Russian ed.) 414:768–71.

Index

T - #0406 - 071024 - C16 - 234/156/12 - PB - 9780367381295 - Gloss Lamination